D1546247

FLUID FLOW POCKET HANDBOOK

Gulf Publishing Company
Book Division
Houston, London, Paris, Tokyo

FLUID FLOW POCKET HANDBOOK

Nicholas P. Cheremisinoff

FLUID FLOW POCKET HANDBOOK

Library of Congress Cataloging in Publication Data

Cheremisinoff, Nicholas P.
    Fluid flow pocket handbook.

    Includes bibliographical references and index.
    1. Fluid Dynamics—Handbooks, manuals, etc. I.
Title.
TA357.C473   1984   620.1'064'0202   83-22619
ISBN 0-87201-707-9

CONTENTS

4

5

6

7

PREFACE

This pocket handbook is intended as a convenient reference to short calculations and design guidelines for fluid flow applications. Its purpose is not to instruct, but rather to summarize and provide a compilation of useful design notes, formulations, and quick computation methods for a wide range of industrial flow problems.

Students, process engineers, and technicians involved with flow problem analysis will find this to be a handy reference. Typical applications of design formulas are illustrated by sample problems and their detailed solutions. A variety of data related to pressure loss and capacity estimation, line sizing, pump selection, pneumatic transport and other frequently encountered industrial situations are included. Sections are prepared in short note form for quick reference and illustration of calculations. Although little or no theory is included, each section refers the user to material that can provide in-depth coverage.

Nicholas P. Cheremisinoff

NOTATION

A	channel flow area
Ar	Archimedes number
a	contraction loss coefficient (bending angle in Chapter 6)
b	channel width
C	cost factor or Chezy discharge coefficient
C_D	discharge coefficient
C'	flow coefficient (see Equation 6-2)
C_e	unit energy cost
C_p, C_v	specific heats at constant pressure and volume, respectively
C_w	nonuniformity flow coefficient
D, d	pipe diameter
D_B	bubble diameter
d_p	particle size
E	energy
e	roughness height or distance
F	cross-sectional area of flow
F'	friction or head loss term
$\hat{F}$	friction energy per unit mass
f	friction factor (cross section in Chapter 10)
G	mass flowrate per unit area
g	gravitational acceleration
g_c	conversion factor
H	fluid head
h_ℓ	head loss or elevation
h_D	hydraulic mean depth
$I_{a,c}$	atomic constants
i	enthalpy
J	mechanical equivalent of heat
j	fractional cost coefficient

K	resistance coefficient (constant in Equation 1-12)
K_c	yearly cost
K_i	yearly expenses for amortization and maintenance
K_p	production and operating expenses
K_s	space factor
K_r	velocity ratio in Equation 8-11
k	flow coefficient for noncircular duct
L	length; liquid mass rate in Chapter 8
ℓ	distance; length of channels in Chapter 9
M	molecular weight
$\tilde{N}$	power
N'	number of tube rows
n	exponent; number of revolutions in Chapter 10
P	pressure
p	perimeter
Q	volumetric flowrate
Q	heat
q	flow per unit width of weir
R	universal gas constant or radius
Re	Reynolds number
r	radius
S_o	orifice cross section
S	slope of fluid free-surface (surface of solids per unit bed volume in Chapter 9)
$\dot{s}$	aspect ratio coefficient
T	absolute temperature
t	temperature (general) or time
t'	distance between tubes
U	fluidization velocity
u	fluid velocity
$U_{L,G}$	superficial liquid and gas velocity, respectively
V	volumetric flow or volume
V_s	superficial velocity
V_m	mixture specific volume
W	mass flowrate
W'	work
W_p, W_s	shaft work
w	velocity
X	cost factor coefficient (function in Chapter 6)
x_i	mass fraction of component i
Y	number of hours of operation per year
y	dummy variable or coordinate
z	elevation

Greek Symbols

α flow regime constant or discharge coefficient; angle of repose in Chapter 9

α' rotameter tube constant

α_o holdup

β poured angle of repose in Chapter 9; pump coefficient in Chapter 10

γ specific weight

$\dot{\gamma}$ shear rate

ϵ relative wall roughness (average void fraction in Chapter 9)

ϵ_j jet constriction coefficient

η efficiency

θ reference temperature

$\varkappa$ ratio of specific heats

λ friction coefficient (two-phase flow parameter in Chapter 8)

μ viscosity

ν kinematic viscosity

ν',ν specific volume

ξ pore volume per unit mass solids

ρ density

σ surface tension

σ_s standard deviation

τ shear stress

ϕ Lockhart-Martinelli parameter

ϕ_s shape factor

χ Lockhart-Martinelli pressure drop parameter

ψ loss coefficient (two-phase flow parameter in Chapter 8)

Subscripts

a   air

B   bubble

c, cr   critical condition

d   discharge

eq   equivalent

f   fluid

fd	fluid discharge
H	homogeneous
h	hydraulic
k	kinetic
ℓ, L	liquid
m	mean
mb	minimum bubbling
mf	minimum fluidization
mot	motor
o	reference
p	particle or potential
r	reduced
s	skeletal
t	turbulent
tr	transmission
v	volumetric
vs	volume surface

1

PHYSICAL
PROPERTIES

Gas Viscosity

For low-pressure pure gases and vapors, the equation of Bromley and Wilke[1] can be used for estimating viscosity:

$$\mu_o = \frac{33.3(M'T_c)^{1/2}}{V_c^{2/3}} [f(1.33\ T_r)] \tag{1-1}$$

where: M' = molecular weight
μ_o = viscosity, micropoise
T_c = critical temperature, °K
V_c = critical volume, cc/g-mole
T_r = reduced temperature, °K

The gas reduced temperature function can be estimated from Scheibel's[2] equation:

$$f(1.33\ T_r) = 1.058\ T_r^{0.645} - \frac{0.261}{(1.9\ T_r)^{0.9\ \log\ (1.9\ T_r)}} \tag{1-2}$$

Estimating procedures for critical and reduced properties of gases are outlined by Perry and Chilton.[3] An alternate method for estimating gas viscosity is illustrated in the following sample calculation.

Sample Calculation 1-1. Estimate the viscosity of sulfur dioxide at a temperature of 300°C and 1 atmosphere pressure. The viscosity of SO_2 at 20°C and 150°C is 1.26×10^{-2} and 1.86×10^{-2} cp, respectively. The critical temperature and pressure are 430°K and 77.7 atm, respectively.

Solution

1. The following equation can be used to estimate viscosity:

$$\mu = 6.3 \times 10^{-4} \frac{M^{1/2} P_{cr}^{2/3}}{T_{cr}^{1/6}} \frac{T_r^{3/2}}{T_r + 0.8} \tag{1-3}$$

where the reduced temperature is defined as $T_r = T/T_{cr}$ and

$$M_{SO_2} = 64; \quad T_r = \frac{(300 + 273)}{430} = 1.335$$

Substituting these values into the previous equation, we obtain the value of viscosity:

$$\mu = (6.3 \times 10^{-4}) \frac{(64)^{1/2} \times (77.7)^{2/3}}{(430)^{1/6}}$$

$$\times \frac{(1.335)^{3/2}}{1.335 + 0.8} = 2.41 \times 10^{-2} cp$$

2. Construct a graph of $y = T^{3/2}/\mu$ versus t using two known values of μ:

$$at \; t_1 = 20°C; \; y_1 = \frac{T^{3/2}}{\mu} = \frac{(239)^{3/2}}{1.26 \times 10^{-2}} = 3.98 \times 10^5$$

$$at \; t_2 = 150°C; \; y_2 = \frac{T^{3/2}}{\mu} = \frac{(423)^{3/2}}{1.86 \times 10^{-2}} = 4.68 \times 10^5$$

The plot is shown in Figure 1-1, from which $y = 5.485 \times 10^5$ at 300°C. Therefore,

$$\mu = \frac{T^{3/2}}{y} = \frac{(573)^{3/2}}{5.485 \times 10^5} = 2.49 \times 10^{-2} cp$$

The same value of y may be obtained using the equation of a straight line passing the points t_1, y_1 and t_2, y_2.

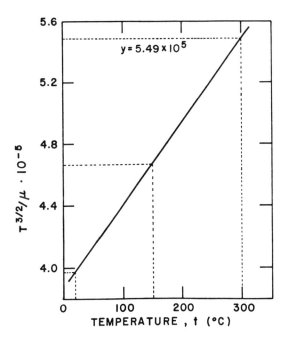

Figure 1-1. Plot of the function $y = T^{3/2}/\mu$ versus temperature.

Alternate Solution. Determine the viscosity at $t = 300°C$, assuming μ_1 at $t = 150°C$ is known. In this case, use the following equation:

$$\mu = \mu_1 \left(\frac{T_r}{T_{r_1}}\right)^{3/2} \times \frac{T_{r_1} + 0.8}{T_r + 0.8} \tag{1-4}$$

$$\mu = 1.86 \times 10^{-2} \left(\frac{573}{423}\right)^{3/2} \times \frac{423/430 + 0.8}{573/430 + 0.8} = 2.44 \times 10^{-2} cp$$

The *effect of temperature* on gas viscosity can be estimated from the following formula:

$$\frac{\mu_2^0}{\mu_1^0} = \left(\frac{T_2}{T_1}\right)^{3/2} \frac{(T_1 + 1.47\ T_b)}{(T_2 + 1.47\ T_r)} \tag{1-5}$$

or

$$\frac{\mu_2^0}{\mu_1^0} = \frac{f(1.33 T_{r_2})}{f(1.33 T_{r_1})} \tag{1-6}$$

The latter equation is preferable for extrapolations over broad temperature ranges.

Caution should be exercised in extrapolating viscosity data to other temperature and pressure conditions. For gases and low-density liquids, empirical methods are best for estimating viscosities. Figure 1-2 shows the method of Uyehara and Watson[4]; consisting of a plot of reduced viscosity versus reduced temperature. Reduced viscosity is defined as the ratio of the viscosity of a given temperature and pressure to the viscosity at the critical point. For low-density gases, viscosity increases with increasing temperature; the opposite is true for liquids. Critical viscosity can be estimated from

$$\mu_c = \frac{61.6(MT_c)^{1/2}}{(V_c)^{-2/3}} \tag{1-7}$$

where μ_c is in micropoise, M is the molecular weight, T_c is in °K, and V_c is the critical molar volume in cc/g-mole.

For *an n-component gas mixture,* viscosity can be estimated from pseudocritical properties and Figure 1-2. The definitions of the pseudocritical properties are:

$$P_c' = \sum_{i=1}^{n} x_i P_{ci} \tag{1-8}$$

$$T_c' = \sum_{i=1}^{n} x_i T_{ci} \tag{1-9}$$

$$\mu_c' = \sum_{i=1}^{n} x_i \mu_{ci} \tag{1-10}$$

Cheremisinoff[5] gives further references and example calculations.

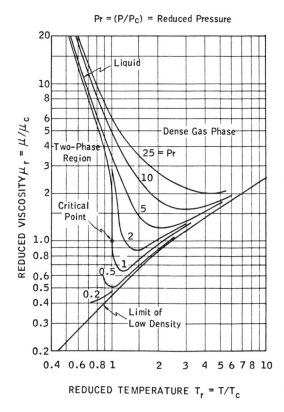

Figure 1-2. Uyehara and Watson's[4] generalized viscosity correlation.

Liquid Viscosity

Table 1-1 summarizes the principal types of viscometers used to measure liquid viscosity and rheological properties of non-Newtonians. Definitions and characteristics of different non-Newtonian fluids are summarized in the Glossary. Detailed discussions on viscometers and non-Newtonian flow properties are given by Skelland[6] and Patel.[7]

Table 1-2 provides typical viscosity ranges and specific gravities of common oils. Table 1-3 provides a conversion chart for viscosity units.

Table 1-1

Major Classifications of Viscometers

Viscometer Type	Capillary	Concentric Cylinder	Cone and Plate
Description			
Measured Variables	Δp–Pressure gradient Q–Fluid flowrate	T–torque on stationary bob Ω–angular velocity of cup	T–torque on stationary plate Ω–angular velocity of cone
Velocity Field	$v_z = v_z(r)$	$v_\theta = v_\theta(r)$	$v_\phi = v_\phi(r, \theta)$
Shear Stress	At tube wall; related to Δp	At bob surface; related to T	At plate; related to T
Shear Rate	At tube wall; related to Q	At bob surface; related to Ω	At plate; related to Ω
Comments	Shear rate varies with r; relation of γ to Q depends on fluid behavior—how Δp varies with Q	Shear rate varies with r; relation of γ to Ω depends on fluid behavior—how T varies with Ω	If angle θ_o is small ($< 1°$), shear rate independent of r; relation of $\dot{\gamma}$ to Ω is independent of fluid behavior
Normal Stress can be obtained from	Axial thrust on capillary; swelling of jet	Difference in pressure at surface of bob and cup	Normal force on cone or plate; pressure distribution across diameter of cone or plate

Table 1-2
Specific Gravity and Viscosity of Common Oils

(Viscosity values reported in Seconds Saybolt Universal Units (SSU); refer to Table 1-3 for conversion to other units.)

LIQUID	SPECIFIC GRAVITY	40° F.	60° F.	80° F.	100° F.	120° F.	140° F.	160° F.
Oils (Transmission)								
SAE 90	.88 -.935	14000	5500	2200	1100	650	380	240
SAE 140	.88 -.935	35000	12000	5000	2200	1200	650	400
SAE 250	.88 -.935	160000	50000	18000	7000	3300	1700	1000
Miscellaneous Oils								
Castor Oil	.96	36000	9000	3000	1400	900	400	300
Chinawood	.943	4000	1800	1000	580	400	300	200
Cocoanut	.925	1500	500	250	140	100	70	60
Cod	.928	1800	600	300	175	110	80	70
Corn	.924	1600	700	400	250	175	100	80
Cotton Seed	.88 -.925	1500	600	300	176	125	80	70
Cylinder	.82 -.95	60000	14000	6000	2700	1400	1000	400
Navy No. 1 Fuel Oil	.989	4000	1100	600	380	200	170	90
Navy No. 2 Fuel Oil	1.0	–	24000	8700	3500	1500	900	480
Gas	.887	180	90	60	50	45	–	–
Insulating		350	150	90	65	50	45	40
Lard	.912-.925	1100	600	380	287	180	140	90
Linseed	.925-.939	1500	500	250	143	110	85	70
Raw Menhadden . . .	.933	1500	500	250	140	110	80	70
Neats Foot	.917	–	1000	430	230	160	100	80
Olive	.912-.918	1500	550	320	200	150	100	80
Palm	.924	1700	700	380	221	160	120	90
Peanut	.920	1200	500	300	195	150	100	80
Quenching	–	2400	900	450	250	180	130	90
Rape Seed	.919	2400	900	450	250	180	130	90
Rosin	.980	28000	7800	3200	1500	900	500	300
Rosin (Wood)	1.09	Extremely Viscose						
Sesame	.923	1100	500	290	184	130	90	60
Soya Bean	.927-.98	1200	475	270	165	120	80	70
Sperm	.883	360	250	170	110	90	70	60
Turbine (Light) . . .	.91	500	350	230	150			
Turbine (Heavy) . . .	.91	3000	1400	700	330	200	150	100
Whale	.925	900	450	275	170	140	100	80
Fuel Oil and Diesel Oil								
No. 1 Fuel Oil	.82 -.95	40	38	35	33	31	30	30
No. 2 Fuel Oil	.82 -.95	70	50	45	40	–	–	-
No. 3 Fuel Oil	.82 -.95	90	68	53	45	40	–	-
No. 5A Fuel Oil	.82 -.95	1000	400	200	100	75	60	40
No. 5B Fuel Oil	.82 -.95	1300	600	490	400	330	290	240
No. 6 Fuel Oil	.82 -.95	–	70000	20000	90000	1900	900	500
No. 2D Diesel Fuel Oil .	.82 -.95	100	68	53	45	40	36	35
No. 3D Diesel Fuel Oil .	.82 -.95	200	120	80	60	50	44	40
No. 4D Diesel Fuel Oil .	.82 -.95	1600	600	280	140	90	68	54
No. 5D Diesel Fuel Oil .	.82 -.95	15000	5000	2000	900	400	260	160

Table 1-3
Conversion Chart for Viscosity Units

Seconds Saybolt Universal ssu	Kinematic Visocity Centistokes	Seconds Saybolt Furol ssf	Seconds Redwood 1 (Standard)	Seconds Redwood 2 (Admiralty)
31	1.00		29	
35	2.56		32.1	
40	4.30		36.2	5.10
50	7.40		44.3	5.83
60	10.3		52.3	6.77
70	13.1	12.95	60.9	7.60
80	15.7	13.70	69.2	8.44
90	18.2	14.4	77.6	9.30
100	20.6	15.24	85.6	10.12
150	32.1	19.30	128	14.48
200	43.2	23.5	170	18.90
250	54.0	28.0	212	23.45
300	65.0	32.5	254	28.0
400	87.60	41.9	338	37.1
500	110.0	51.6	423	46.2
600	132	61.4	508	55.4
700	154	71.1	592	64.6
800	176	81.0	677	73.8
900	198	91.0	462	83.0
1000	220	100.7	896	92.1
1500	330	150	1270	138.2
2000	440	200	1690	184.2
2500	550	250	2120	230
3000	660	300	2540	276
4000	880	400	3380	368
5000	1100	500	4230	461
6000	1320	600	5080	553
7000	1540	700	5920	645
8000	1760	800	6770	737
9000	1980	900	7620	829
10000	2200	1000	8460	921
15000	3300	1500	13700	
20000	4400	2000	18400	

A procedure for estimating the viscosity of pure liquids is outlined in the following sample calculation.

Sample Calculation 1-2. Estimate the visocity of acetic acid at a temperature of 40°C. The density of acetic acid is 1.027 g/cm^3 at 40°C. At 20°C and 100°C, the viscosities are 1.22 cp and 0.46 cp, respectively.

Solution

1. Use the following formula:

$$\log (\log 10\mu) = K\frac{\rho}{M} - 2.9 \qquad (1\text{-}11)$$

where: μ = liquid viscosity, cp
ρ = liquid density, g/cm^3
M = molecular weight
K = constant determined from the following equation:

$$K = \sum mI_a + \sum I_c \qquad (1\text{-}12)$$

where: mI_a = product of the number of atoms, m, of given element in a molecule times its corresponding atomic constant, I_a
I_c = constant characterizing the molecular structure (determined by the atom's grouping and ties among them)

Values of constant I_a for different elements are as follows:

Element	H	C	O	N	Cl	Br	I
Constant I_a	2.7	50.2	29.7	37	60	79	110

Typical values of I_c are given in Table 1-4.

The influence of temperature on liquid viscosity can be determined from viscosity values at two different temperatures by using a comparison with a reference liquid:

$$\frac{t_1 - t_2}{\theta_1 - \theta_2} = K\frac{t_1 - t}{\theta_1 - \theta} \qquad (1\text{-}13)$$

Table 1-4
Values of Ionic Constant I_c[8]

Compound, Grouping or Bond	Constant I_c	Compound, Grouping or Bond	Constant I_c
Double Bond	−15.5		
Five Ring C Atoms	−24.0	CH_3-C-R $\overset{\|}{\underset{O}{}}$	+5.0
Six Ring C Atoms	−21.0		
Substitution in Six-Ring Member In ortho- and para-position	+3.0	$-CH-CHCH_2X^a$	+4.0
In meta-position	+1.0		
$\overset{R \quad\quad R}{\underset{R \quad\quad R}{\diagdown CHCH \diagup}}$	+8.0	$\overset{R}{\underset{R}{\diagup}} CHX^a$	+6.0
$\overset{R}{\underset{R}{R-\overset{\|}{C}-R}}$	+13.0	$-OH$ $-COO-$ $-COOH$ $-NO_2$	+24.7 −19.6 −7.9 −6.4
$R-\overset{\|}{\underset{O}{C}}-H$	+10.0		

[a]Electronegative group.

where: t_1, t_2 = temperature of the liquid
θ_1, θ_2 = temperatures of the reference liquid at which its viscosity is equal to the viscosity of the liquid to be compared at t_1 and t_2

The molecular weight of acetic acid is 60.06. The constant K for acetic acid is

$$K = 2I_a(C) + 4I_a(H) + 2I_a(O) + I_c(COOH)$$
$$= 2 \times 50.2 + 4 \times 2.7 + 2 \times 29.7 + 7.9$$
$$= 162.7$$

Consequently,

$$\log (\log 10\mu) = 162.7 \frac{1.027}{60.06} - 2.9 = -0.118$$

and

$$\log (\log 10\mu) = 0.762; \ \mu = \frac{5.78}{10} = 0.578 \text{ cp}$$

2. Applying Equation 1-13 and using water as the reference liquid, where $\mu'_{H_2O} = 1$, cp and μ''_{H_2O} equals 0.46 cp at $\theta_1 = 12.5$ and θ_2 equals 60.7°C, respectively, the following is obtained:

$$\theta = \theta_1 + (t - t_1)\frac{\theta_2 - \theta_1}{t_2 - t_1} = 12.5$$

$$+ (40\text{-}20) \left(\frac{60.7 - 12.5}{100 - 20} \right) = 24.55°C$$

At $\theta = 24.55°C$, $\mu_{H_2O} = 0.909$ cp, and the viscosity of acetic acid at 40°C is the same.

Other Properties

Principal properties affecting the flow of fluids are density (or specific weight), viscosity, and surface tension. Definitions and useful formulas are given in the Glossary.

Density dependencies on temperature and pressure are described by equations of state. For ideal fluids (gases), the relationship between temperature, pressure, and density is

$$PV = \frac{mRT}{M} \qquad (1\text{-}14)$$

where: P = pressure
       V = gas volume
       m = mass of gas
       R = universal gas law constant (scc Glossary)

T = absolute temperature

M = molecular weight of gas.

In terms of specific volume ($\dot{v} = 1/\rho$), Equation 1-14 can be stated as

$$P\dot{v} = \frac{RT}{M} \tag{1-15}$$

The following example illustrates the use of this equation in SI units.

Sample Calculation 1-3. Compute the density of gaseous ammonia at a pressure of 26 atm. gauge and a temperature of 16°C.

Solution. The absolute pressure is

$$P = 26 + 1 = 27 \text{ kg/cm}^2 = 265 \times 10^4 \text{N/m}^2$$

The molecular weight of ammonia is M = 17. Hence,

$$\rho = PM/RT = \frac{(265 \times 10^4)(17)}{(8314)(273 + 16)} = 1.87 \text{ kg/m}^3$$

Miscellaneous data are given in Tables 1-5 through 1-9. Table 1-5 provides density and specific volume data on water at 1 atm. Table 1-6 gives various properties of air. Table 1-7 gives values for the volume and weight of air at atmospheric pressure as a function of temperature. Table 1-8 provides a convenient chart for evaluating the moisture content in gases in terms of the dew point temperature. Finally, Table 1-9 reports values on the average absolute atmospheric pressure as a function of elevation.

Table 1-5
Density and Volume of Water

Temperature (°C)	(°F)	Density (g/cc)	Volume (cc/g)
−10	14.0	0.99815	1.00186
−8	17.6	0.99869	1.00131
−6	21.2	0.99912	1.00088
−4	24.8	0.99945	1.00055
0	32.0	0.99987	1.00013
4	39.2	1.00000	1.00000
6	42.8	0.99997	1.00003
8	46.4	0.99988	1.00012
10	50.0	0.99973	1.00027
15	59.0	0.99913	1.00087
20	68.0	0.99823	1.00177
25	77.0	0.99708	1.00293
30	86.0	0.99568	1.00434
35	95.0	0.99406	1.00598
40	104.0	0.99225	1.00782
45	113.0	0.99025	1.00985
50	122.0	0.98807	1.01207
55	131.0	0.98573	1.01448
60	140.0	0.98324	1.01705
70	158.0	0.97781	1.02270
80	176.0	0.97183	1.02899
90	194.0	0.96534	1.03590
100	212.0	0.95838	1.04343
110	230.0	0.9510	1.0515
120	248.0	0.9434	1.0601
140	284.0	0.9264	1.0794
160	320.0	0.9075	1.1019
180	356.0	0.8866	1.1279
200	392.0	0.8628	1.1590

Table 1-6
Properties of Air

Air is a mixture of the gases, oxygen and nitrogen, with about 1 per cent by volume of argon. Atmospheric air of ordinary purity contains about 0.04 per cent of carbon dioxide. The composition of air is variously given as follows:

	By Volume			By Weight		
	N	O	Ar	N	O	Ar
1	79.3	20.7	....	77	23	....
2	79.09	20.91	....	76.85	23.15	....
3	78.122	20.941	0.937	75.539	23.024	1.437
4	78.06	21	0.94	75.5	23.2	1.3

The weight of pure air at 32° F. and a barometric pressure of 29.92 inches of mercury, or 14.6963 pounds per square inch, or 2116.3 pounds per square foot, is 0.080728 pound per cubic foot. Volume of 1 pound = 12.387 cubic feet. At any other temperature and barometric pressure its weight in pounds per cubic foot is $W = \frac{1.3253 \times B}{459.6 + T}$ where B = height of the barometer, T = temperature, and 1.3253 = weight in pounds of 459.6 cubic feet of air at 0° F. and 1 inch barometric pressure. Air expands 1/491.6 of its volume at 32° F. for every increase of 1° F., and its volume varies inversely as the pressure.

Table 1-7
Volume and Weight of Air
At Atmospheric Pressure

Temperature Degrees Fahrenheit	Cubic Feet Per Pound	Pounds Per Cubic Foot	Temperature Degrees Fahrenheit	Cubic Feet Per Pound	Pounds Per Cubic Foot
32	12.390	.080710	230	17.379	.057541
50	12.843	.077863	240	17.631	.056718
55	12.969	.077107	250	17.883	.055919
60	13.095	.076365	260	18.135	.055142
65	13.221	.075637	270	18.387	.054386
70	13.347	.074923	280	18.639	.053651
75	13.473	.074223	290	18.891	.052935
80	13.599	.073535	300	19.143	.052238
85	13.725	.072860	320	19.647	.050898
90	13.851	.072197	340	20.151	.049625
95	13.977	.071546	360	20.655	.048414
100	14.103	.070907	380	21.159	.047261
110	14.355	.069662	400	21.663	.046162
120	14.607	.068460	425	22.293	.044857
130	14.859	.067299	450	22.923	.043624
140	15.111	.066177	475	23.554	.042456
150	15.363	.065092	500	24.184	.041350
160	15.615	.064041	525	24.814	.040300
170	15.867	.063024	550	25.444	.039302
180	16.119	.062039	575	26.074	.038352
190	16.371	.061084	600	26.704	.037448
200	16.623	.060158	650	27.964	.035760
210	16.875	.059259	700	29.224	.034219
212	16.925	.059084	750	30.484	.032804
220	17.127	.058388	800	31.744	.031502

Table 1-8
Moisture Content in Gases

To convert parts per million by volume of water vapor to dew points, use the following table:

D.P.	ppm	D.P.	ppm	D.P.	ppm
$-130°F$	0.1	$-73°F$	13.3	$-38°F$	144
-120	0.25	-72	14.3	-37	153
-110	0.63	-71	15.4	-36	164
-105	1.00	-70	16.6	-35	174
-104	1.08	-69	17.9	-34	185
-103	1.18	-68	19.2	-33	196
-102	1.29	-67	20.6	-32	210
-101	1.40	-66	22.1	-31	222
-100	1.53	-65	23.6	-30	235
-99	1.66	-64	25.6	-29	250
-98	1.81	-63	27.5	-28	265
-97	1.96	-62	29.4	-27	283
-96	2.15	-61	31.7	-26	300
-95	2.35	-60	34.0	-25	317
-94	2.54	-59	36.5	-24	338
-93	2.76	-58	39.0	-23	358
-92	3.00	-57	41.8	-22	378
-91	3.28	-56	44.6	-21	400
-90	3.53	-55	48.0	-20	422
-89	3.84	-54	51	-19	448
-88	4.15	-53	55	-18	475
-87	4.50	-52	59	-17	500
-86	4.78	-51	62	-16	530
-85	5.3	-50	67	-15	560
-84	5.7	-49	72	-14	590
-83	6.2	-48	76	-13	630
-82	6.6	-47	82	-12	660
-81	7.2	-46	87	-11	700
-80	7.8	-45	92	-10	740
-79	8.4	-44	98	-9	780
-78	9.1	-43	105	-8	820
-77	9.8	-42	113	-7	870
-76	10.5	-41	119	-6	920
-75	11.4	-40	128	-5	970
-74	12.3	-39	136	-4	1020

Conversion of Parts per Million (ppm) to Per Cent

1 ppm = 0.0001%
10 ppm = 0.001 %
100 ppm = 0.01 %
1,000 ppm = 0.1 %
10,000 ppm = 1.0 %

Table 1-9
Average Absolute Atmospheric Pressure

Altitude in feet referenced to sea level	Inches of mercury (in. Hg)	Pounds per sq. in. absolute (psia)
− 1,000	31.00	15.2
− 500	30.50	15.0
sea level 0	29.92	14.7
+ 500	29.39	14.4
+ 1,000	28.87	14.2
+ 1,500	28.33	13.9
+ 2,000	27.82	13.7
+ 3,000	26.81	13.2
+ 4,000	25.85	12.7
+ 5,000	24.90	12.2
+ 6,000	23.98	11.7
+ 7,000	23.10	11.3
+ 8,000	22.22	10.8
+ 9,000	21.39	10.5
+10,000	20.58	10.1

References

1. Bromley and Wilke, *Ind. Eng. Chem.*, 43, 1641 (1951).
2. Scheibel, E., *Ind. Eng. Chem.*, 46, 1574, 2007 (1954).
3. Perry, R. H., and Chilton, C. H., (eds.), *Chemical Engineer's Handbook, 5th ed.*, McGraw-Hill Book Co., NY (1973), pp. 3.227–3.229.
4. Uyehara, O. A., and Watson, K. M., *Nat. Petrol. Tech.*, Sect. 36, 764, Oct. 4, 1944.
5. Cheremisinoff, N. P., *Fluid Flow: Pumps, Pipes and Channels*, Ann Arbor Science Pub., Ann Arbor, MI (1982).
6. Skelland, A. H. P., *Non-Newtonian Flow and Heat Transfer*, John-Wiley & Sons, Inc., NY (1967).
7. Patel, R. D., pp. 135–177, *Handbook of Fluids In Motion*, N. P. Cheremisinoff, and R. Gupta (editors), Ann Arbor Science Pub., Ann Arbor, MI (1983).
8. Cheremisinoff, N. P. and Azbel, D., *Fluid Mechanics and Unit Operations*, Ann Arbor Science Pub., Ann Arbor, MI (1983).

2

GOVERNING EQUATIONS OF FLOW

Continuity Equation

The amount of fluid flowing through a specified cross section is referred to as the *volumetric flowrate* or *mass (weight) flowrate,* with units of m^3/s or kg/s, respectively. Superficial mass velocity is defined as the ratio of the total mass flowrate and the total area of flow:

$$G = \frac{W}{F}, \ kg/m^2\text{-hr} \ (lb/ft^2\text{-hr}) \tag{2-1}$$

And volumetric flowrate is

$$V = \frac{W}{\rho}, \ m^3/hr \ (ft^3/hr) \tag{2-2}$$

where ρ is fluid density (kg/m^3), and W the mass rate (kg/s).

The mean linear velocity of flow is the ratio of volumetric flowrate to the cross-sectional flow area:

$$\overline{w} = \frac{V}{F}, \ m/hr \ (ft/hr) \tag{2-3}$$

Mass velocity can be expressed as the average velocity and the specific weight of the fluid

$$G = \overline{w} \ \gamma \tag{2-4}$$

where specific weight is the fluid's specific gravity,

$$\gamma = F\frac{G}{V} \tag{2-5}$$

The mass flowrate can be stated as

$$W = \overline{w}\,\gamma\,F \tag{2-6}$$

For a continuous flow system under steady-state conditions, the mass rate must be the same at any section within the system:

$$W_1 = W_2 = W_3 = \dots \tag{2-7}$$

This is the statement of continuity, which can also be written as:

$$G_1F_1 = G_2F_2 = G_3F_3 = \dots \tag{2-8a}$$

$$\overline{w}_1\gamma_1F_1 = \overline{w}_2\gamma_2F_2 = \overline{w}_3\gamma_3F_3 = \dots \tag{2-8b}$$

For flow through a constant cross-sectional area (e.g., a pipe),
$F_1 = F_2 = F$

$$G_1 = G_2 = G_3 = \dots = \overline{w}_1\rho_1 = \overline{w}_2\rho_2 = \overline{w}_3\rho_3 \tag{2-9}$$

If the fluid's properties remain nearly constant from one section to another, then

$$\overline{w}_1F_1 = \overline{w}_2F_2 = \overline{w}_3F_3 \tag{2-10a}$$

or

$$V_1 = V_2 = V_3 = \dots \tag{2-10b}$$

For piping of constant cross section, fluid linear velocities are the same

$$\overline{w}_1 = \overline{w}_2 = \overline{w}_3 = \dots \tag{2-11}$$

The following sample calculation illustrates use of the statement of continuity.

Sample Calculation 2-1. A fluid having specific gravity 0.85 flows through two sections of piping of different cross sections. The average velocity in Section 1 is 1.5 m/s and the pipe diameter is 10 cm. Section 2 of piping has a diameter of 4 cm. Compute: (a) the average velocity in the second pipe section, (b) the volumetric flowrate, (c) the mass flow-rate, (d) the mass velocity in each section.

Solution

1. Applying continuity,

$$\overline{w}_1 F_1 = \overline{w}_2 F_2$$

$$\overline{w}_2 = \overline{w}_1 \times \frac{F_1}{F_2}$$

$$F_1 = \frac{\pi}{4} D_1^2 = \frac{\pi}{4}(0.10m)^2 = 7.85 \times 10^{-3} m^2$$

$$F_2 = \frac{\pi}{4}(0.04m)^2 = 1.26 \times 10^{-3} m^2$$

$$\overline{w}_2 = (1.5 \ m/s)\left(\frac{7.85 \times 10^{-3}}{1.26 \times 10^{-3}}\right) = 9.4 \ m/s$$

2. For the volumetric flowrate,

$$V = F_1 \overline{w}_1 = (7.85 \times 10^{-3} m^2)(1.5 \ m/s) = 1.18 \times 10^{-2} m^3/s$$

or

$$V = F_2 \overline{w}_2 = (1.26 \times 10^{-3} m^2)(9.4 \ m/s) = 1.18 \times 10^{-2} m^3/s$$

3. The mass flowrate is

$$W = V\gamma$$

$$= (1.18 \times 10^{-2} m^3/s)(850 \ kg/m^3) = 10.1 \ kg/s$$

4. The mass velocity is

$$G_1 = \overline{w}_1 \rho = (1.5 \text{ m/s})(850 \text{ kg/m}^3) = 1275 \text{ kg/m}^2\text{-s}$$

$$G_2 = \overline{w}_2 \rho = (9.4 \text{m/s})(850 \text{ kg/s}) = 7990 \text{ kg/m}^2\text{-s}$$

Flow Regimes

Three general flow regimes are observed in flow systems: (1) laminar, (2) transition, and (3) turbulent. The laminar regime occurs at relatively low fluid velocities. The flow in this case may be viewed as layers that slide over each other providing smooth flow patterns. No macroscopic mixing of fluid particles occurs.

In the turbulent regime fluid velocities are higher, and an unstable pattern within the bulk flow is observed in which eddy currents move at all angles to the axis of normal flow. The transition regime denotes the onset of turbulence.

The dependency of flow regime is summarized by the dimensionless Reynolds number:

$$Re = \frac{WD}{\nu} = \frac{wD\rho}{\mu} = \frac{WD}{\mu} \tag{2-12}$$

where: w = fluid velocity
W = mass velocity
D = system or tube diameter
μ = dynamic viscosity
ρ = density
ν = fluid kinematic viscosity ($= \mu/\rho$)

The critical Reynolds number denotes the transition from one regime to another. For a straight circular pipe when the Reynolds number is less than 2100, the flow is laminar. When the Reynolds number exceeds 4000, the flow is turbulent. The flow between these two critical values is transitional.

For laminar flow in a tube, the volumetric flowrate can be computed from the Poiseuille equation (see Cheremisinoff[1,2] for derivation):

$$V = \frac{\pi D^2 \Delta P}{128 \mu L} \tag{2-13}$$

where L is the pipe length, and ΔP the pressure drop over the length of the pipe.

Or, in terms of tube radius R

$$\overline{w} = \frac{\Delta P}{8\mu L} R^2 \qquad (2-14)$$

The average velocity for laminar flow in a circular tube is approximately half the maximum velocity at the pipe axis:

$$\overline{w} = \frac{w_{max}}{2} \qquad (2-15)$$

The parabolic velocity distribution for pipe flow is

$$w_r = 2\overline{w}\left(1 - \frac{r^2}{R^2}\right) \qquad (2-16)$$

Total Energy Balance

Consider the flow system shown in Figure 2-1. The energy input of the system is the sum of the kinetic E_{k1}, potential E_{p1}, volumetric E_{v1}, and internal E_{i1} energies at section 1; the heat Q' added through the exchanger; and the mechanical work W' performed on the fluid by the pump. The total energy output consists of kinetic E_{k2}, potential E_{p2}, volumetric E_{v2}, and internal E_{i2} energies at section 2:

$$E_{k1} + E_{p1} + E_{v1} + Q' + W' = E_{k2} + E_{p2} + E_{v2} + E_2 \qquad (2-17)$$

Potential energy is defined as the product of weight and the fluid's elevation with respect to a specified datum (for 1 kg of weight):

$$E_{p1} = z_1; \; E_{p2} = z_2$$

The volumetric energy under pressure P is equivalent to the work expended to form volume v' at this pressure. The volumetric energies of 1 kg of fluid at the two sections are:

$$E_{v1} = P_1 v'_1; \; E_{v2} = P_2 v'_2 \qquad (2-18)$$

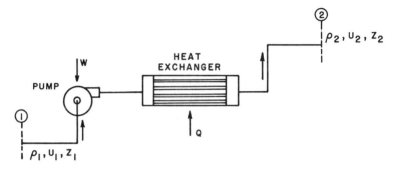

Figure 2-1. General flow system for explanation of the total energy balance.

Kinetic energy is the product of mass and one-half the square of the fluid's linear velocity. The weight of the fluid (i.e., the product of fluid mass and gravitational acceleration g) is equal to 1 kilogram-force (kgf). The mass of the fluid in gravitational units is $1/g \times kg \times s^2/m^4$. For velocity w, kinetic energy is $w^2/2g$

$$E_{k1} = \frac{\overline{w}_1^2}{2g\alpha}; \; E_{k2} = \frac{\overline{w}_2^2}{2g\alpha_2} \tag{2-19}$$

α is a correction coefficient. For turbulent flow, $\alpha = 1$.

Internal energy is a thermodynamic property of the flow system; defined relative to some specified reference state (e.g., 0°C and 1 atm)

$$P_1 v_1' + \frac{\overline{w}_1^2}{2g\alpha_1} + z_1 + Q + W'$$
$$= P_2 v_2' + \frac{\overline{w}_2^2}{2g\alpha_2} + z_2 + (E_2 - E_1) \tag{2-20}$$

Equation 2-20 is the law of conservation of energy. For engineering calculations, the internal energy terms may be approximated by the enthalpies:

$$i_1 = E_1 + P_1 v_1'; \; i_2 = E_2 + P_2 v_2' \tag{2-21}$$

Hence,

$$\frac{\overline{w}_1^2}{2g\alpha_1} + z_1 + Q + W' = \frac{\overline{w}_2^2}{2g\alpha_2} + z_2 + (i_2 - i_1) \qquad (2\text{-}22)$$

For ideal gases, the change in enthalpy can be computed from the product of heat capacity at constant pressure C_p and the system's temperature difference, $C_p(t_2 - t_1)$.

Bernoulli Equation

The Bernoulli equation is a special form of the total energy balance and can be stated as follows:

$$z_1 + \frac{P_1}{\rho g} + \frac{w_1^2}{2\alpha g} = z_2 + \frac{P_2}{\rho g} + \frac{w_2^2}{2\alpha g} \qquad (2\text{-}23)$$

The hydrodynamic head $(2 + P/\rho g + w^2/2g)$ is constant for all cross sections of ideal steady flow.

Note that z is referred to as the leveling height (also called the geometric head). $P/\rho g$ is the static or piezometric head characterizing the specific potential energy of pressure at a given point. Both terms may be expressed in units of length or specific energy (i.e., energy per unit weight of fluid).

The term $w^2/2g$ is also in units of length:

$$\frac{\overline{w}_2}{2g} \equiv \left[\frac{m^2 \times s^2}{s^2 \times m}\right] \equiv [m]$$

and is called the velocity or dynamic head. It characterizes the specific kinetic energy at a given point within the flow cross section.

For real (viscous) fluids, the Bernoulli equation must be corrected for frictional losses (referred to as the *lost head* h_ℓ):

$$z_1 + \frac{P_1}{\rho g} + \frac{\overline{w}_1^2}{2g} = z_2 + \frac{P_2}{\rho g} + \frac{\overline{w}_1^2}{2g} + h_\ell \qquad (2\text{-}24)$$

or

$$\rho g z_1 + \frac{\rho \overline{w}_1^2}{2\alpha} + P_1 = \rho g z_2 + P_2 + \frac{\rho \overline{w}_2^2}{2\alpha} + \Delta P_\ell \qquad (2\text{-}25)$$

where ΔP_ℓ is the lost pressure drop

$$\Delta P_\ell = \rho g h_\ell \qquad (2\text{-}26)$$

The following sample calculation illustrates use of the Bernoulli equation.

Sample Calculation 2.2. A fluid having a specific gravity of 0.9 is being pumped through a constant area pipe. The pressure upstream of the pump is 1.2 lb_f/ft^2 abs, and the pump's discharge pressure is 3.5 lb_f/ft^2. The discharge line on the pump is 26 feet above the centerline of the inlet. The pump supplies 61 $ft\text{-}lb_f/lb_m$ of fluid through the pipe. The flow through the pipe system is in the turbulent regime. Determine the frictional losses in the system.

Solution. The centerline of the pump inlet is chosen as the reference datum. Hence, $z_1 = 0$ and $z_2 = 26$ feet.
Rewriting Equation 2-25 and solving for the friction losses,

$$\sum h_\ell = -\hat{W}_s + \frac{1}{2\alpha}(w_1^2 - w_2^2) + g(z_1 - z_2) + \frac{P_1 - P_2}{\rho}$$

$\hat{W}_s$ is the work performed by the pump, and Σh_ℓ is the sum of the friction losses (i.e., head losses).
Evaluating each term,

$$\frac{1}{2(1)}(w_1^2 - w_2^2) = 0$$

$$g(z_1 - z_2) = (32.2 \text{ ft/s}^2)(0 - 26 \text{ ft}) = -837 \text{ ft}^2/\text{s}^2$$

$$g_c \times \frac{P_1 - P_2}{\rho} = \frac{(1.2 - 3.5)[lb_f/ft^2]}{\left(0.9 \times 62.4 \dfrac{lb_m}{ft^3}\right)}$$

$$\times 32.2 \frac{ft\text{-}lb_m}{lb_f\text{-}s^2} = -1.32 \text{ ft}^2/\text{s}^2$$

Hence,

$$\sum h_\ell = -\left(-61\,\frac{\text{ft-lb}_f}{\text{lb}_m}\right)\left(32.2\,\frac{\text{ft-lb}_m}{\text{lb}_f\text{-s}^2}\right) + 0$$

$$+ \left(-837\,\frac{\text{ft}^2}{\text{s}^2}\right) + \left(-1.32\,\frac{\text{ft}^2}{\text{s}^2}\right) = 1126\,\frac{\text{ft}^2}{\text{s}^2}$$

or

$$\sum h_\rho = 1126\,\frac{\text{ft}^2}{\text{s}^2} \times \frac{\text{lb}_f\text{-s}^2}{32.2\;\text{ft-lb}_m} = 34.9\,\frac{\text{ft-lb}_f}{\text{lb}_m}$$

Sample Calculation 2-3. A liquid (density = 73 lbs/ft³) is being pumped at a rate of 45 gpm from an open tank to a height of 40 feet above the initial level in the vessel. The discharge line is 2.5-in. ID, and the total friction loss in the piping is 23 ft-lb$_f$/lb$_m$. The level in the tank falls at a rate of 0.25 fps and the pump's rated efficiency is 62%. Compute the pump's horsepower. Flow in the lines is turbulent.

Solution. The mechanical energy delivered to the fluid by the pump is

$$\hat{W} = -\eta W_p$$

where: η = fractional pump efficiency
W_p = shaft work delivered to the pump

The volumetric flowrate is

$$V = \left(45\,\frac{\text{gal}}{\text{min}}\right)\left(\frac{1\;\text{min}}{60\;\text{s}}\right)\left(\frac{\text{ft}^3}{7.481\;\text{gal}}\right) = 0.1003\;\text{cfs}$$

The fluid velocity in the tank is given as w_1 = 0.25 fps. The cross section of the discharge pipe is

$$F = \frac{1}{4}\pi(2.5/12)^2 = 0.0341\;\text{ft}^2$$

Hence, the velocity downstream of the pump is

$$w_2 = \frac{0.1003\;\text{cfs}}{0.0341\;\text{ft}^2} = 2.940\;\text{fps}$$

We will assume that the discharge is open to the atmosphere. Hence,

$$P_1 = P_2 = 1 \text{ atm}$$

Therefore,

$$\frac{P_1}{\rho} - \frac{P_2}{\rho} = 0$$

And because $\alpha = 1.0$ (i.e., turbulent flow conditions),

$$\frac{w_1^2}{2g_c} = \frac{(0.25 \text{ fps})^2}{2(32.174)} = 9.713 \times 10^{-4} \frac{\text{ft-lb}_f}{\text{lb}_m}$$

$$\frac{w_2^2}{2g_c} = \frac{(2.940 \text{ fps})^2}{2(32.174)} = 0.1343 \frac{\text{ft-lb}_f}{\text{lb}_m}$$

Assigning the initial level in the tank as the reference datum plane,

$$z_1 = 0$$

and

$$z_2 \frac{g}{g_c} = (40 \text{ ft})\left(\frac{32.2}{32.174}\right) = 40.0 \frac{\text{ft-lb}_f}{\text{lb}_m}$$

Rearranging the mechanical energy equation to solve for mechanical work,

$$\hat{W}_s = \frac{g}{g_c}(z_1 - z_2) + \frac{1}{2g_c}(w_1^2 - w_2^2) + \frac{P_1 - P_2}{\rho} - \sum \hat{F}$$

$$\hat{W}_s = 40 + (9.713 \times 10^{-4} - 0.1343) + 0 - 23$$

$$= -63.13 \frac{\text{ft-lb}_f}{\text{lb}_m}$$

Hence,

$$\hat{W}_p = \frac{-\hat{W}_s}{\eta} = \frac{-(-63.13)}{0.62} = 101.8 \frac{\text{ft-lb}_f}{\text{lb}_m}$$

Mass flowrate $= (0.1003 \text{ cfs})(73 \text{ lbs}_m/\text{ft}^3) = 7.32 \text{ lbs}_m/\text{s}$. The pump horsepower is therefore

$$\left(7.32 \frac{\text{lb}_m}{\text{s}}\right)\left(101.8 \frac{\text{ft-lb}_f}{\text{lb}_m}\right)\left(\frac{1 \text{ hp}}{550 \frac{\text{ft-lb}_f}{\text{s}}}\right) = 1.36 \text{ hp}$$

References

1. Cheremisinoff, N. P., and Azbel, D. S., *Fluid Mechanics and Unit Operations,* Ann Arbor Science Pub., Ann Arbor, MI (1983).
2. Cheremisinoff, N. P., *Fluid Flow: Pumps, Pipes and Channels,* Ann Arbor Science Pub., Ann Arbor, MI (1981).

3

PIPE
FLOW
CALCULATIONS

Frictional Pressure Losses

This section outlines methods for evaluating pressure losses through piping and related equipment for liquid flows.

The pressure drop in a horizontal straight length of pipe of constant diameter is caused by friction and can be calculated from the Fanning friction equation. Fanning friction factor f is a function of the Reynolds number and relative pipe wall roughness and is shown in Figure 3-1. For a given class of pipe material, roughness is relatively independent of pipe diameter; therefore, the friction factor can be expressed as a function of Reynolds number and pipe diameter. For laminar flow (Re < 2000), the friction factor is independent of pipe wall roughness and can be expressed as a function of Reynolds number alone.

The accuracy of the Fanning friction equation is $\pm 15\%$ for smooth tubing and $\pm 10\%$ for commercial steel pipe. Fouling can reduce the cross-sectional area or increase pipe wall roughness with time. Therefore, when calculating pressure drop, one should allow for fouling.

The majority of data concerning the effect of fouling on pressure drop are for water piping. For such piping, instead of the Fanning correlation, an empirical correlation known as the Hazen-Williams correlation has been most widely used. The correlation contains a coefficient known as the Hazen-Williams "C" factor, which is used to account for surface condition and fouling.

In piping designs one of the primary requirements is to determine an inside diameter that will allow a certain required throughput at a given pressure drop. This involves a trial-and-error procedure. A diameter is chosen, and the pressure drop is computed for the required throughput. If the calculated pressure drop is too great, a larger diameter is chosen

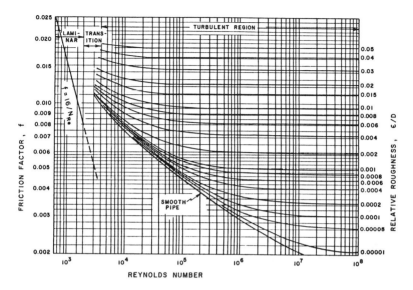

Figure 3-1. Friction factor plot for any type of commercial pipe.

for the next trial. If the pressure drop is smaller than necessary, a smaller diameter is chosen.

The basic equation for calculating pressure drop for liquid flow in pipes and fittings is the generalized Bernoulli equation, which assumes constant density:

$$-\frac{10^{-3}\Delta P}{\rho g} = \frac{\alpha\Delta(w^2)}{2g} + \Delta z + \frac{10^{-3}F'}{F'} \qquad (3\text{-}1)$$

Pressure change	Kinetic energy change	Eleva-tion change	Friction or head loss

where the units in SI are as follows:

F' = friction or head loss, kPa-m^3/kg
g = acceleration of gravity 9.81 m/s
ΔP = pressure change, kPa
w = velocity of the fluid, m/s

z = elevation, m

ρ = density, kg/m^3

α = constant depending on velocity profile (α = 1.0 for turbulent flow, α = 2.0 for laminar flow).

The relative importance of the terms in the equation varies from application to application, but general definitions are given in the previous section. For constant-diameter horizontal pipes, only the friction term on the right-hand side of the equation is important. For vertical or inclined pipes, one must include the elevation term; for cross-sectional changes, the kinetic energy term.

For Newtonian liquids, both constant viscosity and density can be assumed. Non-Newtonian liquids are an exception to this rule. Another exception is nonisothermal flow due either to heat exchange or to heat production or consumption in the liquid by chemical reaction or friction losses.

Where the flow may be assumed to be isothermal across the pipe cross section but not isothermal along the length of the pipe, the pressure drop can be determined by dividing the pipe into a number of lengths and calculating the pressure drop in each section.

The effects of bends, tees, valves, orifices, and other flow restrictions is to cause additional pressure drop in pipelines. Fittings that have the same nominal diameter as the pipe can be accounted for in terms of an *equivalent length* of straight pipe. This equivalent length can be computed from the resistance coefficients of the fittings (typical values of which are given in Table 3-1). The equivalent length is then added to the actual length of the pipe, and the sum is used in the Fanning equation for predicting the total friction pressure drop.

When piping details are not available, the following guidelines may be used for estimating equivalent length.

For *onsite lines* the actual pipe length can be estimated from the plot plan, tower heights, etc. Equivalent length of fittings in onsite piping adds between 200% and 500% to the actual length. Consequently, a multiplier of 3.0–6.0 may be applied to the estimated length of straight pipe.

For *offsite lines* the approximate length of straight pipe can be estimated from the plot plan. Since fittings in offsite lines usually have an equivalent length of 20–80% of the actual length, a multiplier of 1.2–1.8 can be applied to the estimated length of straight pipe.

The pressure drop in cross-sectional changes, such as exits and entrances of process vessels and reducers and diffusers, consists of two components: one for friction and one for variations in kinetic energy.

Table 3-1
Loss Coefficients for Turbulent Flow Through Valves and Fittings[1]

Type Fitting or Valve	Head Loss Coefficient (K_e)	L_e/D
Elbow, 45°	0.35	17
Elbow, 90°	0.75	35
Tee	1	50
Return Bend	1.5	75
Coupling	0.04	2
Union	0.04	2
Gate Valve		
Wide open	0.17	9
Half open	4.5	225
Glove Valve		
Wide open	6.0	300
Half open	9.5	475
Angle Valve		
Wide open	2.0	100
Check Valve		
Ball	70.0	3500
Swing	2.0	100
Water Meter, disk	7.0	350

Calculation of the friction loss is based on the diameter of the smaller of the two pipes with no obstructions.

For pipes ending in an area of very large cross section, such as a process vessel, the frictional pressure drop is equal to the gain in pressure caused by the change in kinetic energy. The net pressure change over the cross-sectional change is zero.

For a very gradual contraction, friction pressure drop is calculated based on a straight piece of pipe with inside diameter equal to the narrowest cross section of the contraction.

In pressure drop calculations for lines containing fittings and cross-sectional changes, the line is first broken into sections of constant nominal diameter. The frictional pressure drop of each change in cross section is accounted for in the equivalent length of the smaller diameter pipe attached to it. The pressure drop due to the various changes in kinetic energy in the line is determined by computing the overall change in kinetic energy between the inlet and outlet of the line.

When a stream is split into two or more substreams, there is both a friction loss and a pressure change due to the change in kinetic energy. The same applies to the combining of streams. For tees, the total pressure change is given by the equations listed in Table 3-2.

Equations and guidelines for estimating pressure losses in piping systems are outlined in the following paragraphs. Several examples are given to illustrate the procedures for calculating pressure drop in single piping components and in flow systems containing more than one piping component.

Examples of single piping components are runs of straight pipe, bends, valves, orifices, etc. If the pipe has a noncircular cross section, first compute the equivalent hydraulic diameter:

$$d_{eq} = 4\left(\frac{\text{cross-sectional area}}{\text{wetted perimeter}}\right) \tag{3-2}$$

For a given diameter and flow rate, compute the Reynolds number Re from the following equation:

$$Re = \frac{Dw\rho}{\mu} = 10^{-3}\left(\frac{Dw\rho}{\mu}\right)$$

$$= 1.27\left(\frac{Q\rho}{d\mu}\right) \tag{3-3}$$

$$= 1.27 \times 10^{3}\left(\frac{W}{d\mu}\right)$$

where: D = ID of pipe or equivalent hydraulic diameter, m
d = ID of pipe or equivalent hydraulic diameter, mm
Q = volumetric flow rate, dm^3/s
Re = Reynolds number, dimensionless
w = velocity, m/s
W = mass flow rate, kg/s
ρ = density, kg/m^3
μ = dynamic viscosity, Pa-s

Table 3-2
Pressure Drop Equations for Split and Join Flow Streams[2]
(All units are given in SI)

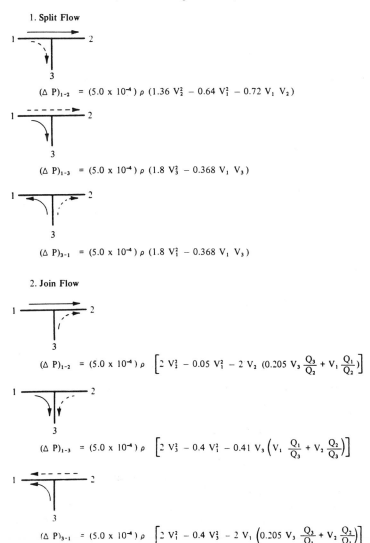

1. **Split Flow**

$$(\Delta P)_{1-2} = (5.0 \times 10^{-4})\, \rho\, (1.36\, V_2^2 - 0.64\, V_1^2 - 0.72\, V_1\, V_2)$$

$$(\Delta P)_{1-3} = (5.0 \times 10^{-4})\, \rho\, (1.8\, V_3^2 - 0.368\, V_1\, V_3)$$

$$(\Delta P)_{3-1} = (5.0 \times 10^{-4})\, \rho\, (1.8\, V_1^2 - 0.368\, V_1\, V_3)$$

2. **Join Flow**

$$(\Delta P)_{1-2} = (5.0 \times 10^{-4})\, \rho\, \left[2\, V_2^2 - 0.05\, V_1^2 - 2\, V_2\, \left(0.205\, V_3\, \frac{Q_3}{Q_2} + V_1\, \frac{Q_1}{Q_2}\right)\right]$$

$$(\Delta P)_{1-3} = (5.0 \times 10^{-4})\, \rho\, \left[2\, V_3^2 - 0.4\, V_1^2 - 0.41\, V_3\, \left(V_1\, \frac{Q_1}{Q_3} + V_2\, \frac{Q_2}{Q_3}\right)\right]$$

$$(\Delta P)_{3-1} = (5.0 \times 10^{-4})\, \rho\, \left[2\, V_1^2 - 0.4\, V_3^2 - 2\, V_1\, \left(0.205\, V_3\, \frac{Q_3}{Q_1} + V_2\, \frac{Q_2}{Q_1}\right)\right]$$

Figure 3-1 should then be used to obtain a value of friction factor f. For values of Re lower than those covered by this figure, with Re < 2000 (laminar flow), calculate f from the following:

$$f = \frac{16}{Re} \tag{3-4}$$

where f is the dimensionless friction factor.

Next, compute the frictional pressure drop from the following equations:

$$(\Delta P)_f = 10^{-3}\left(\frac{4fL}{D}\right)\left(\frac{\rho w^2}{2}\right)$$

$$= 2\frac{fLw^2\rho}{d}$$

$$= 3.24 \times 10^6\left(\frac{fLQ^2\rho}{d^5}\right) \tag{3-5}$$

$$= 3.24 \times 10^{12}\left(\frac{fLW^2}{\rho d^5}\right)$$

where: $(\Delta P)_f$ = frictional pressure drop, kPa
         L = pipe length, m

If the pipe is not horizontal, the pressure drop due to the change in elevation must be computed:

$$(\Delta P)_e = 10^{-3}(\rho g)(z_2 - z_1) \tag{3-6}$$

where: $(\Delta P)_e$ = pressure drop due to change in elevation, kPa
         z_1, z_2 = elevation of beginning and end of pipe, m.

As in the previous section, the total pressure drop is obtained by adding the frictional pressure drop $(\Delta P)_f$ and the pressure drop due to change in elevation $(\Delta P)_e$. Resistance coefficients for bends, ells and tees are given in Figure 3-2.

For pipes larger than 250-mm ID, use the resistance coefficient for 250-mm ID pipe. If the Reynolds number is such that the flow is not in the region of complete turbulence (f is constant), the value of K should be multiplied by the ratio:

$$\frac{f_{\text{(at calculated Reynolds number)}}}{f_{\text{(in range of complete turbulence)}}}$$

Using the resistance coefficients from Figure 3-2, compute the frictional pressure drop from the following equation:

$$(\Delta P)_f \; = \; 10^{-3}\left(\frac{K\rho w^2}{2}\right) \tag{3-7}$$

For long nonhorizontal bends, add the pressure drop due to the change in elevation calculated from Equation 3-6.

For blanked-off tees and Y's use Equation 3-7 and the resistance coefficients for tees in Figure 3-2. For tees in which streams are split or joined, the pressure drop should be calculated from the equations given in Table 3-2.

The equations given in Table 3-2 account for both frictional pressure drop and pressure drop due to changes in kinetic energy. To account for entrance and exit effects in cases where the inlet leading line is short, a multiplying factor of 1.25 can be applied.

Sample Calculation 3-1. Compute the pressure loss across a one-half-open gate valve located in a horizontal 6-in. Schedule 40 pipe carrying 400 gpm of SAE 10W oil at 92°F.

Solution. Applying the Bernoulli equation,

$$\frac{P_1}{\gamma} + z_1 + \frac{w_1^2}{2g} - h_\ell = \frac{P_2}{\gamma} + z_2 + \frac{w_2^2}{2g}$$

Since this is a horizontal system,

$$z_1 \; = \; z_2 \; = \; 0$$

and

$$P_1 - P_2 \; = \; \gamma\left(0 + \frac{w_2^2 - w_1^2}{2g} + h_\ell\right)$$

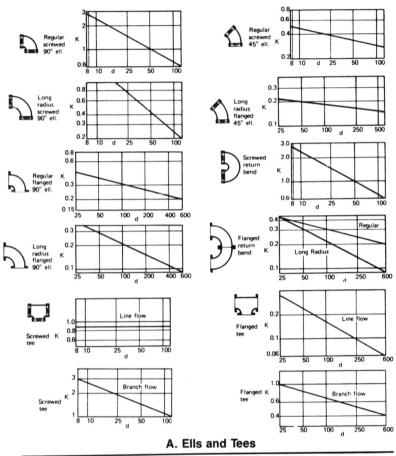

A. Ells and Tees

(Figure 3-2 Continued on next page)

At steady-state, $w_1 = w_2$; hence,

$$P_1 - P_2 = \rho g h_\ell$$

The head loss can be computed from the following:

$$h_\ell = f\left(\frac{\Sigma L_e}{D}\right)\frac{w^2}{2g}$$

Smooth pipe bends

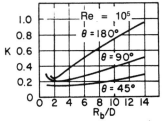

Reynolds no.	Multiplier for K
10^4	1.48
10^5	1.00
10^6	0.676

Length of pipe in bend is included in K as additional loss. Elsewhere, length contribution is excluded from K.

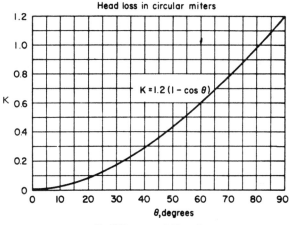

Head loss in circular miters

$K = 1.2\,(1 - \cos\theta)$

θ, degrees

B. Miters and Bends

Figure 3-2. Resistance coefficients for bends, ells, and tees.[3]

From Table 3-1, for a half-open gate valve, $L_e/D = 225$. The actual diameter of a 6-in. Schedule 40 pipe is 0.5054 feet. Hence,

$$F = \frac{\pi}{4}(0.5054)^2 = 0.201 \ \text{ft}^2$$

$$w = \frac{400 \ \text{gpm}}{0.201 \ \text{ft}^2} \times \frac{1 \ \text{ft}^3/\text{s}}{449 \ \text{gpm}} = 4.4 \ \text{fps}$$

The kinematic viscosity of the fluid is $\nu = 1.8 \times 10^{-4} \text{ft}^2/\text{s}$. The Reynolds number is

$$Re = \frac{wD}{\nu} = \frac{(4.4)(0.5054)}{1.8 \times 10^{-4}} = 12{,}445$$

The wall roughness is $e = 1.3 \times 10^{-4}$ inches; hence,

$$\epsilon = D/e = \frac{0.5054}{1.2 \times 10^{-4}} = 4{,}212$$

From Figure 3-1, the friction factor is obtained ($f = 0.031$). Hence,

$$h_\ell = (0.031)(225)\frac{(4.4)^2}{2(32.2)} = 2.096 \text{ ft}$$

$$P_1 - P_2 = \gamma h_\ell = (0.875)\left(62.4\frac{\text{lb}}{\text{ft}^3}\right)(2.096 \text{ ft})\left(\frac{\text{ft}^2}{144 \text{ in}^2}\right)$$

$$= 0.795 \text{ psi}$$

Flow Through Varying Cross Sections

If the flow enounters a change in cross section, additional head losses result. For a sudden contraction, losses that arise may be estimated from the Weisbach equation:

$$h_\ell = (0.04 + (1/a - 1)^2)\frac{w_2^2}{2g} = \psi\frac{w_2^2}{2g} \qquad (3\text{-}8)$$

where a is the contraction loss coefficient, defined as the ratio of the minimum flow cross section to the cross section of the smaller pipe; and w_2 is the average velocity in the smaller section.

Coefficients a and ψ depend on the ratio of the pipe's cross sections. Typical values are given as follows:

F_1/F_2	0.01	0.1	0.2	0.4	0.6	0.8	1.0
a	0.6	0.61	0.62	0.65	0.7	0.77	1.0
ψ	0.5	0.46	0.42	0.33	0.23	0.13	0.0

Coefficient ψ can also be estimated from the following formula:

$$\psi = \frac{1.5(1 - F_2/F_1)}{3 - F_2/F_1} \qquad (3-9)$$

For a sudden expansion, an expression can be derived from the Bernoulli equation for turbulent flow (see Cheremisinoff[1]). An approximate formula is

$$h_\ell = \frac{(w_1 - w_2)^2}{2g} \qquad (3-10)$$

Sample Calculation 3-2. Water is flowing at a rate of 2500 gal/hr through a distribution system. At one point, the pipeline suddenly expands from a 1-inch diameter to 1.5 inches. Compute (a) the energy loss that occurs due to the sudden enlargement, and (b) the difference between the pressure immediately ahead of the sudden enlargement and downstream.

Solution

$F_1 = \pi/4(1/12 \text{ ft})^2 = 0.00545 \text{ ft}^2$
$F_2 = \pi/2(1.5/12 \text{ ft})^2 = 0.01670 \text{ ft}^2$

Hence,

$$w_1 = Q/F_1 = \frac{2500 \text{ gph}}{0.00545 \text{ ft}^2} \times \frac{1 \text{ ft}^3/\text{s}}{449 \text{ gpm}} \times \frac{\text{hr}}{60 \text{ min}} = 17.0 \text{ fps}$$

$$w_2 = \frac{2500}{0.01670} \times \frac{1}{449} \times \frac{1}{60} = 5.56 \text{ fps}$$

$$h_\ell = \frac{(w_1 - w_2)^2}{2g} = \frac{(17.0 - 5.56)^2}{2(32.2)} = 2.03 \text{ ft}$$

That is, 2.03 ft-lb$_f$ of energy is dissipated for each lb$_m$ of fluid flowing through the sudden enlargement.

Gradual expansions and contractions are illustrated in Figure 3-3. For smooth conical expansions (for $7° < \beta < 35°$, see Figure 3-3(a)),

$$h_\ell = 0.35\left(\log \frac{\beta}{2}\right)^{1.22} \frac{(w_1 - w_2)^2}{2g} \qquad (3-11)$$

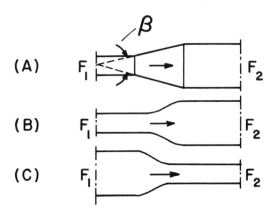

Figure 3-3. Flow through gradual expansions and contractions.

At $\beta > 40°$, head losses may be very high and even exceed those in sudden expansions.

For gradual contractions as in Figure 3-3c, head losses are very small. Equation 3-8 can be used allowing $h_\ell = 0.05$, independent of the ratio of F_2/F_1, provided that the flow is turbulent in the narrow cross section. If the flow in the contraction section is laminar, a pressure decrease generally is observed that does not follow Poiseuille's law.

This decrease occurs over a length equivalent to 0.065 ReD; the entrance region of the pipe. The pressure gradient at the entrance of a pipe of length L can be estimated from the following data:

$\dfrac{L}{D}$Re	0.005	0.01	0.02	0.03	0.04	0.05	0.06
$\dfrac{p_0 - p_1}{\gamma \dfrac{w^2}{2g}}$	2.1	2.6	3.4	4.1	4.7	5.3	6.0

[a]D = pipe diameter; L = pipe length; γ = liquid specific gravity.

An alternate approach to estimating friction losses for gradual expansions with turbulent flow is to use the following equation:

$$h_\ell = C_L \frac{(w_1 - w_2)^2}{2g} \qquad (3\text{-}12)$$

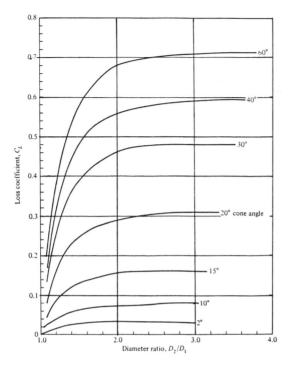

Figure 3-4. Loss coefficients for gradual expansions.[4,5]

Values of the loss coefficient C_L as a function of diameter ratios and different cone angles are given in Figure 3-4. The following sample calculations illustrate both methods.

A similar calculation can be made for conical diffusers (refer to Figure 3-5) and for sudden enlargements and contractions (Figure 3-6). The following sample calculations illustrate both methods.

Sample Calculation 3.3. Compute the energy loss for 25 gpm of water flowing from a 1-in. tube into a 2.9-in. tube via a gradual enlargement with an included angle of 30°.

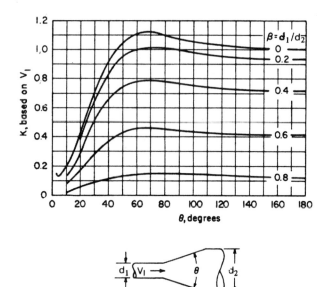

Figure 3-5. Head loss coefficients in conical diffusers.[4,5]

Solution

$$Q = 25 \text{ gpm} \times \frac{ft^3}{7.48 \text{ gal}} \times \frac{min}{60 \text{ s}} = 0.0557 \text{ ft}^3/s$$

$$F_1 = \frac{\pi}{4}(1/12)^2 = 0.00545 \text{ ft}^2; \; w_1 = \frac{0.0557}{0.00545} = 10.2 \text{ fps}$$

$$F_2 = \frac{\pi}{4}(2.9/12)^2 = 0.0459 \text{ ft}^2; \; w_2 = \frac{0.0557}{.0459} = 1.21 \text{ fps}$$

$$h_\ell = 0.35 \left(\log \frac{\beta}{2}\right)^{1.22} \frac{(w_1 - w_2)^2}{2g}$$

$$= 0.35 \; (\log 15)^{1.22} \frac{(10.2 - 1.21)^2}{2 \times 52.2} = 0.54 \text{ ft (or 0.24 psi)}$$

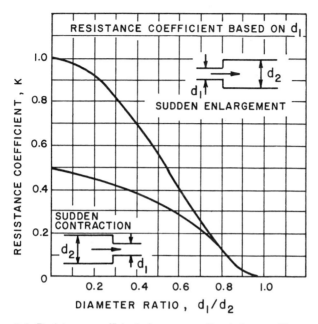

Figure 3-6. Resistance coefficients for cross-sectional changes.[4,5]

Alternate Solution

$D_2/D_1 = 2.9/1 = 2.9$

From Figure 3-4 for $\beta = 30°$, $C_L \simeq 0.47$

$$h_\ell = C_L \frac{(w_1 - w_2)^2}{2g}$$

$$= 0.47 \frac{(10.2 - 1.21)^2}{2 \times 32.2} = 0.56 \text{ ft (or 0.25 psi)}$$

Sample Calculation 3-4. A water holding tank is drained from a 1.5-in. ID side nozzle. Estimate the energy loss as the flow undergoes a sudden contraction. The discharge rate is 40 gpm.

Solution

$$h_\ell = \left[0.04 + \left(\frac{1}{a} - 1\right)^2\right]\frac{w_2^2}{2g} = \psi\frac{w_2^2}{2g}$$

$$\psi = 1.5(1 - F_2/F_1)/(3 - F_2/F_1) \simeq 0.5 \text{ (since } F_1 >> F_2)$$

$$Q = 40 \text{ gpm} \times \text{cfs}/449 \text{ gpm} = 0.0891 \text{ cfs}$$

$$F_2 = \frac{\pi}{4}(1.5/12)^2 = 0.0123 \text{ ft}^2;$$

$$w_2 = 0.0891/0.0123 = 7.24 \text{ fps}$$

$$h_\ell = 0.5\frac{(7.24)^2}{2 \times 32.2} = 0.407 \text{ ft (or 0.179 psi)}$$

Efflux from Vessels and Pipes

Efflux from an open vessel involves direct application of Bernoulli's equation. Consider the system in Figure 3-7 where the vessel is open, with constant level. Hence, $P_1 = P_2$ and $w_1 = 0$. Neglecting the distance between the orifice plane at the tank floor and plane 2-2, $z_1 - z_2 = H$

$$\frac{w_2^2}{2g} = H$$

and

$$w_2 = \sqrt{2gH}$$

A jet velocity factor is applied to this expression to account for head lost to friction and for overcoming the resistance due to jet constriction in the orifice

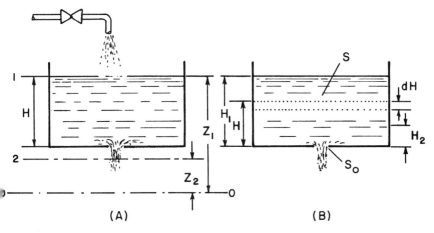

Figure 3-7. Efflux from a tank: (a) constant liquid level, and (b) varying liquid level.

$$w_2 = \psi\sqrt{2gH} \tag{3-13}$$

Since the jet cross section in the orifice S_o is larger than its narrowest section S_2, the velocity in the orifice w_o must be less than w_2

$$w_o = \epsilon_j w_2 = \epsilon_j \psi\sqrt{2gH} = \alpha\sqrt{2gH} \tag{3-14}$$

where: $\epsilon_j = S_2/S_o$ (the jet constriction coefficient)
  $\alpha = $ discharge coefficient $ = \psi\epsilon_j$

Coefficient α must be experimentally determined. For liquids similar to water, $\alpha = 0.62$. The volumetric rate is

$$Q = \alpha S_o \sqrt{2gH} \tag{3-15}$$

The previous expressions are also valid for liquid discharging through an orifice in the tank's side wall. In this case, the level H should be measured from the orifice centerline.

Figure 3-7b shows efflux from a variable level tank. An expression for estimating the time of discharge from a vessel of constant cross section is (see Cheremisinoff et al.[1] for derivation)

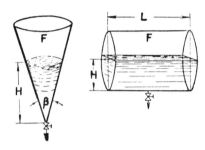

Figure 3-8. Liquid discharge from variable cross-section vessels.

$$t = \frac{2S}{\alpha S_o \sqrt{2g}} (\sqrt{H_1} - \sqrt{H_2}) \tag{3-16}$$

If the vessel is emptied completely, then $H_2 = 0$ and

$$t = \frac{2S \sqrt{H_1}}{\alpha S_o \sqrt{2g}} \tag{3-17}$$

Figure 3-8 illustrates two other common vessel geometries. To determine the discharge times from such vessels, the relationship between cross section and height must be known. For the conical vessel, the cross section as a function of height H is

$$S = \pi H^2 \tan^2 \beta/2 \tag{3-18}$$

For the horizontal cylindrical vessel,

$$S = 2L\sqrt{HD - H^2} \tag{3-19}$$

The discharge times for these vessels can be estimated by substituting the appropriate expression for S into Equation 3-16.

For discharges from partially filled pipes, the following empirical formula of Folsom[6] is recommended:

$$Q = 2.54 D^{2.56} K_s^{1.84} \tag{3-20}$$

where Q is in units of ℓ/s and pipe diameter D in m. K_s is the space factor, defined as the ratio of fluid level (normal from the pipe floor) to pipe diameter. Equation 3-20 is applicable over the ranges of K = $0.2 \sim 0.6$ and D = $3 \sim 15$ cm.

Optimum Pipe Diameter

The optimum pipe size is defined as that diameter which provides an acceptable resistance per unit length within specified economic constraints. As noted earlier, this involves a trial-and-error solution.

Several pipe sizes will be determined acceptable in terms of head losses for a given flow rate. However, selecting a larger diameter leads to higher construction costs and, over a planned operating lifetime, maintenance and repair and amortization expenses K_i will also be high. At the same time, a large diameter means lower hydraulic resistance and, consequently, lower power consumption for transportation; hence, production expenses K_p will be relatively low. Conversely, with a smaller diameter, hydraulic resistance increases along with production costs, whereas the amortization expenses decrease. The relationship of K_i (maintenance/amortization costs) and K_p (production costs) to pipe diameter are shown in Figure 3-9. The sum of K_i and K_p is the total yearly expenses for a pipe network. This is also shown in Figure 3-9 as a plot of ΣK versus diameter. The relationship has a minimum which corresponds to the optimum pipe diameter.

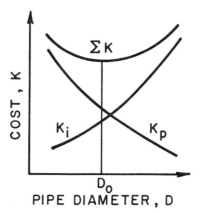

Figure 3-9. Relationship between operating costs and capital investment to pipe size.

Determination of the optimum pipe diameter may be formulated into a rigorous design procedure which is outlined here. Both series of expenses should be related to a unit length of piping and a basis of one year of operation. The cost per unit length of piping is directly proportional to the diameter

$$C_1 = XD^n \tag{3-21}$$

From data supplied by manufacturers on the cost of piping and installation, coefficients X and n can be evaluated.

Pipe components such as fittings and valves are also included in the analysis. The cost of fittings depends on the corresponding pipe diameter and may be considered a fractional cost of the piping (i.e., $C_2 = jC$, where j is a coefficient denoting fractional cost). The total cost per unit length C of a piping network, including fittings and valves, is

$$C = (1 + j)XD^n \tag{3-22}$$

where $C = C_1 + C_2$.

By assigning an operating life for the system (e.g., 10–15 years) one can evaluate the portions of the total cost C attributed to amortization expenses (denote as a) and to maintenance costs (b), such as repairs, painting, insulation, etc. That is, the total yearly expenses for amortization and maintenance are

$$K_i = (a + b)(1 + j)XD^n \tag{3-23}$$

Production or operating expenses depend on the number of hours of operation per year (Y), the hourly rate of energy needed for fluid transportation $(\tilde{N})$, and the cost of a unit of energy (C_e):

$$K_p = Y\tilde{N}C_e \tag{3-24}$$

The power required for fluid transportation may be expressed as $Q\Delta P/\eta$, where Q is the volumetric rate, ΔP is the pressure drop per unit length of pipe, and η is the mechanical pump efficiency. Hence,

$$K_p = \frac{W_g\Delta pKC_e}{\gamma\eta} \tag{3-25}$$

where W_g is the weight rate.

From the Darcy equation for a unit length of pipe

$$\Delta P = \frac{8}{\pi^2} \frac{\lambda w^2}{g\gamma D^5} \tag{3-26}$$

For a Reynolds number range of 4000 to 2×10^7, the following equation may be used to obtain a friction factor:

$$\lambda = \frac{0.16}{Re^{0.16}} \tag{3-27}$$

where:

$$Re = \frac{GD}{\mu g} \tag{3-28}$$

$$G = \frac{W}{\pi D^2/4}$$

Hence, the friction factor expression is

$$\lambda = \frac{0.16}{\left(\dfrac{w}{\pi D^2/4} \times \dfrac{D}{\mu g}\right)^{0.16}} = 0.154 \left(\frac{D\mu g}{w}\right)^{0.16} \tag{3-29}$$

The previous equations are combined (see Cheremisinoff[1] for detailed derivation) to get

$$K_p = 0.125 \frac{w^{2.84}\mu^{0.16}YC_e}{D^{4.84}\gamma^2\eta g^{0.84}} \tag{3-30}$$

Equations 3-30 and 3-23 provide estimates of production costs and amortization and maintenance expenses, respectively. The sum of these two is the total yearly cost for a piping system

$$\Sigma K_c = (a + b)(j + 1)XD^n + 0.125 \frac{w^{2.84}\mu^{0.16}YC_e}{D^{4.84}\gamma^2\eta^2 g^{0.84}} \tag{3-31}$$

The optimum diameter is the minimum cost incurred for the pipe system. This minimum cost can be obtained by differentiating Equation 3-31 with respect to D and setting the derivative equal to zero; the result is the expression for optimum diameter

$$D_0^{4.84} = \frac{0.605w^{2.84}\mu^{0.16}YC_e}{n(a + b)(j + 1)X\gamma^2\eta g^{0.84}} \tag{3-32}$$

For pipe diameters greater than $3/4$ inch, exponent n in Equation 3-21 is nearly unity. The final expression for the optimum pipe diameter is

$$D_0 = 0.918\left[\frac{YC_e}{(a + b)(j + 1)X\eta}\right]^{0.17}\frac{\mu^{0.027}}{g^{0.14}}\frac{w^{0.48}}{\gamma^{0.34}} \tag{3-33}$$

Examining this expression reveals that the dependency of D_0 on viscosity is very small ($\mu^{0.027}$). For example, in the range of 0.02 cp (the viscosity of air at STP) to 30 cp (a typical oil), $\mu^{0.027}$ changes by only a few percent. Hence, $\mu^{0.027}$ can be assumed as a constant. Denoting

$$K_1 = 0.918\left[\frac{YC_e}{(a + b)(j + 1)X\eta}\right]^{0.17}\frac{\mu^{0.027}}{g^{0.14}} \tag{3-34}$$

as a constant (actually assuming K_1 to be constant is a good approximation to within 10%), the optimum pipe diameter expression may be written simply as:

$$D_0 = K_1\frac{w^{0.48}}{\gamma^{0.34}} \tag{3-35}$$

The derivation of Equation 5-21 is based on the condition of turbulent flow. Hence, the D_0 calculation should be checked by computing the Reynolds number and comparing it against the criterion for turbulent pipe flow.

A similar design procedure is given by Normand[7] for laminar flows:

$$D_0^{4+n} = \frac{512}{\pi}\frac{w^2YC_e\mu}{\gamma^2\eta n(a + b)(j + 1)X} \tag{3-36}$$

For $D_0 > 3/4$-inch, the exponent n equals 1, and the expression simplifies to

$$D_0 = 2.77\left[\frac{YC_e}{(a + b)(j + 1)X\eta}\right]^{0.2}\left(\frac{w^2\mu}{\gamma^2}\right)^{0.2} \tag{3-37}$$

Denoting

$$K_2 = 2.77 \left[\frac{YC_e}{(a + b)(j + 1)X\eta} \right] \tag{3-38}$$

as a constant, the following expression is obtained for calculating the optimum pipe diameter for laminar flow:

$$D_0 = K_2 \left(\frac{w^2 \mu}{\gamma^2} \right)^{0.2} \tag{3-39}$$

For laminar flow $Re = GD_0/\mu g < 2100$, or substituting $w/\pi D^2/4$ for G,

$$Re = wD_0 \left/ \frac{\pi D_0^2}{4} \mu g \right. < 2100 \tag{3-40}$$

Replacing D_0 in this expression with Equation 3-39 gives

$$W_g < 2.2 \times 10^5 (K_2 g)^{1.67} \frac{\mu^2}{\gamma^{0.67}} = W_{gcr} \tag{3-41}$$

This last expression evaluates the flow regime that will exist for the optimum pipe diameter. Hence, from information on the weight rate W_g, viscosity μ, specific weight γ, and computed constant K_2 (from Equation 3-38), the proper expression for calculating the optimum pipe diameter (Equations 3-35 or 3-39) can be selected. The following sample calculation illustrates the design procedure.

Sample Calculation 3-5. Compute the optimum size pipe for transporting 5000 kg/hr of water. The minimum life of the pipeline may be assumed to be 10 years, and annual maintenance and repair expenses are estimated at 5% of the initial piping costs. The cost of fittings and valves is roughly 10% of the cost of the pipe. The system is to be designed for 24-hour operation throughout. The pump efficiency, η, is 60%, and the cost of electrical power is 0.2 $/kW-hr. The following data provide quotations on costs per unit length of piping for different pipe diameters:

D (in.)	$3/4$	1	$1^1/4$	$1^1/2$	2	$2^1/2$	3	$3^1/2$	4	5	6
D (mm)	21.5	27.0	35.75	41.25	52.5	68	80.25	92.5	105	130	155.5
Price ($/m)	3.12	4.30	4.94	5.80	7.51	9.04	11.30	13.7	15.4	19.9	23.7

The cost per unit length of piping is given by Equation 3-21 $C_1 = XD^n$, where $n \simeq 1$. Plotting the quotation data given and evaluating the slope, we find $X \simeq 150$ \$/m². Hence $C_1 \simeq 150$ D(\$/m)

Electrical costs are

$$C_e = 0.02 \text{ \$/kW/hr} = \frac{0.2}{3.67 \times 10^5} = 5.45 \times 10^{-7} \text{ \$/kg-m}$$

Operating time is

$$Y = 365 \times 24 \times 3600 = 3.15 \times 10^7 \text{ sec/year}$$

The amortization expenses are $a = 0.10$. Maintenance and repair costs are 5% of the pipe costs (or $b = 0.05$). Hence,

$$a + b = 0.10 + 0.05 = 0.15$$

The cost of all fittings is 10% of the capital costs for piping. Hence, $j = 0.1$.

From these values, the term common to Equations 3-34, 3-37 and 3-41 may be evaluated:

$$\frac{YC_e}{(a + b)(j + 1)X\eta} = \frac{(3.15 \times 10^7)5.45 \times 10^{-7}}{0.15 \times 1.1 \times 150 \times 0.6} = 1.20$$

Using Equation 3-41, we evaluate the transition from laminar to turbulent flow:

$$\mu = \frac{1}{9.8 \times 10^3} = 1.02 \times 10^{-4} \frac{\text{kg-s}}{\text{m}^2}$$

$$K_2 = 2.77(1.20)^{0.2} = 2.9$$

Hence, the critical mass velocity for flow through the optimum pipe size is

$$W_{g,cr} = 2.2 \times 10^5(2.9 \times 9.81)^{1.67}\frac{(1.02 \times 10^{-4})}{1000^{0.67}}$$

$$= 7.35 \times 10^3 \text{kg/sec}$$

Table 3-3
Areas of Circles in Square Inches

Diam. inches	0	1/8"	1/4"	3/8"	1/2"	5/8"	3/4"	7/8"
0	....	.01227	.04909	.1104	.1963	.3068	.4418	.6013
1	.7854	.9940	1.2272	1.4849	1.7671	2.0739	2.4053	2.7612
2	3.1416	3.5466	3.9761	4.4301	4.9088	5.4119	5.9396	6.4918
3	7.0686	7.6699	8.2958	8.9462	9.6212	10.3206	11.0447	11.7933
4	12.566	13.364	14.186	15.033	15.904	16.800	17.721	18.665
5	19.635	20.629	21.648	22.691	23.758	24.850	25.967	27.109
6	28.274	29.465	30.680	31.919	33.183	34.472	35.785	37.122
7	38.485	39.871	41.282	42.718	44.179	45.664	47.173	48.707
8	50.266	51.849	53.456	55.088	56.745	58.426	60.132	61.862
9	63.617	65.397	67.201	69.029	70.882	72.760	74.662	76.589
10	78.540	80.516	82.516	84.541	86.590	88.664	90.763	92.886
11	95.033	97.205	99.402	101.623	103.869	106.139	108.434	110.753
12	113.10	115.47	117.86	120.28	122.72	125.19	127.68	130.19
13	132.73	135.30	137.89	140.50	143.14	145.80	148.49	151.20
14	153.94	156.70	159.48	162.30	165.13	167.99	170.87	173.78
15	176.71	179.67	182.65	185.66	188.69	191.75	194.83	197.93
16	201.06	204.22	207.39	210.60	213.82	217.08	220.35	223.65
17	226.98	230.33	233.71	237.10	240.53	243.98	247.45	250.95
18	254.47	258.02	261.59	265.18	268.80	272.45	276.12	279.81
19	283.53	287.27	291.04	294.83	298.65	302.49	306.35	310.24
20	314.16	318.10	322.06	326.05	330.06	334.10	338.16	342.25
21	346.36	350.50	354.66	358.84	363.05	367.28	371.54	375.83
22	380.13	384.46	388.82	393.20	397.61	402.04	406.49	410.97
23.	415.48	420.00	424.56	429.13	433.74	438.36	443.01	447.69
24	452.39	457.11	461.86	466.64	471.44	476.26	481.11	485.98

For this problem, $W_g = 5000/3600 = 1.39$ kg/sec, i.e., the flow through the optimum piping is turbulent. Since $W_{g,cr} > W_g$, the flow is turbulent and Equation 3-35 should be used to compute D_0.

From Equation 3-34,

$$K_1 = 0.918(1.20)^{0.17} \frac{(1.02 \times 10^{-4})^{0.027}}{9.81^{0.14}} = 0.545$$

Finally, from Equation 3-35,

$$D_0 = 0.545 \frac{1.39^{0.48}}{(1000)^{0.34}} = 6.1 \times 10^{-2} m$$

Pipe and Pipe Flow Data

This section contains miscellaneous data useful to pipe flow calculations. Table 3-3 provides cross-sectional areas of circles in units of

Table 3-4
Equivalent Length of Schedule 40 Pipe (in Feet)
Which Must Be Added to the Length of Run
to Obtain Total Pressure Drop

Nominal Size of Pipe in Inches	Using Cast Iron Drainage Fittings	
	90° Change in Direction	45° Change in Direction
1-1/4	3	1-1/2
1-1/2	4	2
2	5	2-1/2
2-1/2	6	3
3	7	4
4	10	5
5	12	6
6	15	7-1/2
8	20	10

square inches. Table 3-4 provides equivalent lengths of Schedule 40 pipe for pressure drop calculations. Table 3-5 provides pressure drop data for water flowing through Schedule 40 steel pipe under standard conditions. Table 3-6 provides dimensions of steel pipe.

Table 3-5
Pressure Drop of Water Through
Schedule 40 Steel Pipe

Based on Saph and Schoder Formulas $\Delta P = \dfrac{LQ^{1.86}}{1435\, d^{5}}$

FLOW — **Pressure Drop of Water per 100 Ft. of Schedule 40 Steel Pipe in psi**

In the table below, each pipe-size cell is given as **v (Ft/Sec) / p (psi)**.

G.P.M.	FT³ Per Sec.	1/8″	1/4″	3/8″	1/2″	3/4″	1″	1 1/4″	1 1/2″	2″	2 1/2″	3″	3 1/2″	4″	5″	6″	8″	10″	12″	14″	16″	18″
.1	.00022	56 / .677																				
.2	.00045	1.14 / 2.48	.62 / .548																			
.3	.00067	1.70 / 5.26	.93 / 1.16	.50 / .255																		
.4	.0089	2.26 / 9.00	1.24 / 1.98	.67 / .436																		
.5	.00111	2.82 / 13.58	1.55 / 3.00	.84 / .656	.42 / .136																	
.6	.00134	3.38 / 19.12	1.85 / 4.22	1.01 / .925	.53 / .205																	
.8	.00178	4.52 / 32.62	2.47 / 7.17	1.34 / 1.58	.84 / .494																	
1	.00223		3.09 / 10.91	1.68 / 2.39	1.06 / .749	.60 / .183	.37 / .055	.21 / .014														
2	.00446		6.18 / 39.60	3.36 / 8.68	2.11 / 2.72	1.20 / .665	.74 / .199	.43 / .051														
3	.00668			5.04 / 18.46	3.17 / 5.77	1.80 / 1.41	1.11 / .424	.64 / .107														
4	.00891			6.72 / 31.55	4.22 / 9.86	2.40 / 2.42	1.49 / .724	.86 / .183														
5	.01114				5.28 / 14.92	3.01 / 3.64	1.86 / 1.09	1.07 / .276														
6	.01337				6.33 / 20.95	3.61 / 5.13	2.23 / 1.54	1.29 / .390	1.26 / .308													
8	.01782					4.81 / 8.76	2.97 / 2.62	1.71 / .667	1.58 / .466													
10	.02228					6.01 / 13.28	3.713 / 3.97	2.142 / 1.01														
15	.03342						5.57 / 8.46	3.21 / 2.14	2.36 / .992	1.43 / .285												
20	.04456						7.43 / 14.42	4.28 / 3.66	3.15 / 1.69	1.91 / .486												
25	.05570							5.36 / 5.54	3.94 / 2.54	2.39 / .736	2.01 / .424											
30	.06684							6.43 / 7.79	4.73 / 3.60		2.35 / .566											
35	.07798							7.50 / 10.38	5.51 / 4.79	3.35 / 1.37	2.68 / .724											
40	.08912							8.57 / 13.28	6.30 / 6.14	3.82 / 1.76												
50	.1114								7.88 / 9.31	4.78 / 2.66	3.35 / 1.10	2.17 / .371										
60	.1337								9.45 / 13.08	5.74 / 3.75	4.02 / 1.54	2.61 / .520	2.27 / .335									
70	.1560									6.70 / 4.99	4.69 / 2.05	3.04 / .693	2.59 / .430									
80	.1782									7.65 / 6.40	5.37 / 2.63	3.47 / .890	2.92 / .535									
90	.2005									8.60 / 7.96	6.04 / 3.28	3.91 / 1.10	3.24 / .650									
100	.2228									9.56 / 9.69	6.71 / 3.98	4.34 / 1.34	3.24 / .650	2.52 / .346								
125	.2785										8.38 / 6.03	5.43 / 2.01	4.05 / .984	3.15 / .523								
150	.3342										10.1 / 8.46	6.52 / 2.86	4.87 / 1.38	3.78 / .734								
175	.3899										11.7 / 11.3	7.60 / 3.81	5.68 / 1.84	4.41 / .978	2.81 / .316							
200	.4456										13.4 / 14.4	8.69 / 4.89	6.49 / 2.36	5.04 / 1.25	3.21 / .405							
225	.5013											9.77 / 6.09	7.30 / 2.94	5.67 / 1.56	3.61 / .505							
250	.5570											10.9 / 7.41	8.11 / 3.58	6.30 / 1.90	4.01 / .616	2.78 / .245						
275	.6127											11.9 / 8.84	8.92 / 4.27	6.93 / 2.27	4.41 / .734	3.06 / .292						
300	.6684											13.0 / 10.4	9.73 / 5.02	7.56 / 2.67	4.81 / .863	3.33 / .344						
350	.7798											15.2 / 13.8	11.4 / 6.87	8.82 / 3.55	5.62 / 1.15	3.89 / .457						
400	.8912												13.0 / 8.58	10.1 / 4.56	6.41 / 1.47	4.44 / .587	2.57 / .149					
450	1.003												14.6 / 10.7	11.3 / 5.66	7.22 / 1.83	5.00 / .731	2.89 / .185					
500	1.114												16.2 / 13.0	12.6 / 6.89	8.02 / 2.23	5.55 / .887	3.21 / .225					
550	1.225												17.8 / 15.5	13.9 / 8.25	8.82 / 2.67	6.11 / 1.07	3.53 / .270					
600	1.337												19.5 / 18.2	15.2 / 9.68	9.62 / 3.13	6.66 / 1.25	3.85 / .316					
650	1.449													16.4 / 11.2	10.4 / 3.62	7.22 / 1.45	4.17 / .367	2.65 / .118				
700	1.560													17.6 / 12.9	11.2 / 4.16	7.78 / 1.66	4.49 / .420	2.85 / .135				
750	1.671													18.9 / 14.7	12.0 / 4.75	8.33 / 1.89	4.81 / .480	3.05 / .154				
800	1.782													20.2 / 16.5	12.8 / 5.35	8.89 / 2.13	5.13 / .540	3.26 / .173				
850	1.894													21.4 / 18.5	13.6 / 5.98	9.44 / 2.38	5.45 / .605	3.46 / .194				
900	2.005													22.7 / 20.6	14.4 / 6.65	10.0 / 2.66	5.77 / .627	3.66 / .216	2.58 / .090			
950	2.117													23.9 / 22.8	15.2 / 7.36	10.6 / 2.93	6.09 / .744	3.87 / .238	2.72 / .099			
1000	2.228														16.0 / 8.10	11.1 / 3.23	6.41 / .817	4.07 / .262	2.87 / .109			
1100	2.451														17.6 / 9.66	12.2 / 3.85	7.06 / .975	4.48 / .313	3.15 / .130	2.85 / .096		
1200	2.674														19.2 / 11.4	13.3 / 4.53	7.70 / 1.15	4.88 / .368	3.44 / .153	3.08 / .111		
1300	2.896														20.8 / 13.2	14.4 / 5.26	8.34 / 1.33	5.29 / .427	3.73 / .178	3.32 / .127		
1400	3.119														22.4 / 15.1	15.6 / 6.01	8.98 / 1.53	5.70 / .490	4.01 / .204	3.56 / .145		
1500	3.342														24.1 / 17.2	16.7 / 6.84	9.62 / 1.74	6.10 / .556	4.30 / .232	3.79 / .163	2.91 / .084	
1600	3.565															17.8 / 7.73	10.3 / 1.96	6.51 / .628	4.59 / .262		3.27 / .104	
1800	4.010															20.0 / 9.64	11.5 / 2.46	7.32 / .782	5.16 / .329	4.27 / .203	3.63 / .127	
2000	4.456															22.2 / 11.6	12.8 / 2.97	8.14 / .953	5.73 / .396	4.74 / .247	3.63 / .127	
2500	5.570															27.8 / 17.6	16.0 / 4.49	10.2 / 1.44	7.17 / .601	5.93 / .374	4.54 / .192	4.30 / .149
3000	6.684																19.2 / 6.30	12.2 / 2.02	8.60 / .842	7.11 / .525	5.45 / .270	5.02 / .199
3500	7.798																22.4 / 8.41	14.2 / 2.70	10.0 / 1.12	8.30 / .700	6.36 / .358	5.74 / .255
4000	8.912																25.7 / 10.8	16.3 / 3.31	11.5 / 1.44	9.48 / .896	7.26 / .459	6.45 / .317
4500	10.03																28.9 / 13.4	18.3 / 4.31	12.9 / 1.76	10.7 / 1.12	8.17 / .571	7.17 / .386
5000	11.14																	20.4 / 5.20	14.3 / 2.18	11.9 / 1.36	9.08 / .695	8.17 / .386
6000	13.37																	24.4 / 7.35	17.2 / 3.06	14.2 / 1.91	10.9 / .977	8.60 / .542
7000	15.60																	28.5 / 9.80	20.1 / 4.08	16.6 / 2.54	12.7 / 1.30	10.0 / .723
8000	17.82																		22.9 / 5.22	19.0 / 3.25	14.5 / 1.67	11.5 / .926
9000	20.05																		25.8 / 6.51	21.3 / 4.06	16.3 / 2.08	12.9 / 1.15
10000	22.28																		28.7 / 7.91	23.7 / 4.92	18.2 / 2.53	14.3 / 1.40
12000	26.74																			28.5 / 6.92	21.8 / 3.55	17.2 / 1.97
14000	31.19																				25.4 / 4.72	20.1 / 2.62
16000	35.65																				29.1 / 6.06	22.9 / 3.36
18000	40.10																				32.7 / 7.55	25.8 / 4.18
20000	44.56																					28.7 / 5.08

Table 3-6
Dimensions of Steel Pipe

Nominal Pipe Size (in.)	OD (in.)	Schedule No.	ID (in.)	Flow Area per Pipe (in.^{2})	Surface per lin ft (ft^2/ft) Outside	Inside	Weight per lin ft lb Steel
1/8	0.405	40[a]	0.269	0.058	0.106	0.070	0.25
		80[b]	0.215	0.036		0.056	0.32
1/4	0.540	40[a]	0.364	0.104	0.141	0.095	0.43
		80[b]	0.302	0.072		0.079	0.54
3/8	0.675	40[a]	0.493	0.192	0.177	0.129	0.57
		80[b]	0.423	0.141		0.111	0.74
1/2	0.840	40[a]	0.622	0.304	0.220	0.163	0.85
		80[b]	0.546	0.235		0.143	1.09
3/4	1.05	40[a]	0.824	0.534	0.275	0.216	1.13
		80[b]	0.742	0.432		0.194	1.48
1	1.32	40[a]	1.049	0.864	0.344	0.274	1.68
		80[b]	0.957	0.718		0.250	2.17
1 1/4	1.66	40[a]	1.380	1.50	0.435	0.362	2.28
		80[b]	1.278	1.28		0.335	3.00
1 1/2	1.90	40[a]	1.610	2.04	0.498	0.422	2.72
		80[b]	1.500	1.76		0.393	3.64
2	2.38	40[a]	2.067	3.35	0.622	0.542	3.66
		80[b]	1.939	2.95		0.508	5.03
2 1/2	2.88	40[a]	2.469	4.79	0.753	0.647	5.80
		80[b]	2.323	4.23		0.609	7.67
3	3.50	40[a]	3.068	7.38	0.917	0.804	7.58
		80[b]	2.900	6.61		0.760	10.3
4	4.50	40[a]	4.026	12.7	1.178	1.055	10.8
		80[b]	3.826	11.5		1.002	15.0
6	6.625	40[a]	6.065	28.9	1.734	1.590	19.0
		80[b]	5.761	26.1		1.510	28.6
8	8.625	40[a]	7.981	50.0	2.258	2.090	28.6
		80[b]	7.625	45.7		2.000	43.4
10	10.75	40[a]	10.02	78.8	2.814	2.62	40.5
		60	9.75	74.6		2.55	54.8
12	12.75	30	12.09	115	3.338	3.17	43.8
14	14.0	30	13.25	138	3.665	3.47	54.6
16	16.0	30	15.25	183	4.189	4.00	62.6
18	18.0	20[c]	17.25	234	4.712	4.52	72.7
20	20.0	20	19.25	291	5.236	5.05	78.6
22	22.0	20[c]	21.25	355	5.747	5.56	84.0
24	24.0	20	23.25	425	6.283	6.09	94.7

[a]Commonly known as standard.
[b]Commonly known as extra heavy.
[c]Approximately.

References

1. Cheremisinoff, N. P. and Azbel, D. S., *Fluid Mechanics and Unit Operations,* Ann Arbor Science Pub., Ann Arbor, MI (1983).
2. Cheremisinoff, N. P., *Fluid Flow: Pumps, Pipes and Channels,* Ann Arbor Science Pub., Ann Arbor, MI (1982).
3. *Chemical Engineering,* 75, No. 13, 198–199 (June 17, 1968).
4. Crane Co., Technical Paper No. 410, *Flow Through Valves, Fittings and Pipe* (1970).
5. Simpson, L. L., "Process Piping: Functional Design," *Chem. Eng.,* 76, No. 8, (Deskbook Issue), 167–181 (April 14, 1969).
6. Folsom, R. A., Trans. Am. Soc. Mech. Eng., 78, 1447–1460 (1956).
7. Normand, C. E., *Ind. Eng. Chem.,* 40(5):783 (1948).

4

GAS
FLOW
CALCULATIONS

Basis of Pressure Drop Calculations

The flow of gases through piping is more complex than liquids because of the dependency of specific weight on pressure changes. Because gases undergo thermodynamic changes with pressure, densities vary greatly.

The basic equations for computing pressure drop for the flow of gases through piping systems are based on the steady-state energy balance and the differential form of the Bernoulli equation:

$$J\Delta E + \frac{g}{g_c}\Delta Z + 144\Delta(P\nu) + \frac{\Delta(w^2)}{2g_c} = J\dot{Q} - W_s \qquad (4\text{-}1)$$

$$\frac{g}{g_c}dz + \nu\,144\,dP + \frac{w\,dw}{g_c} = -dF' - dW_s \qquad (4\text{-}2)$$

where: E = internal energy, Btu/lb_m
F' = frictional energy loss, ft lb_f/lb_m
g = acceleration of gravity, ft/s^2
g_c = dimensional constant, 32.174 ft lb_m/lb_f-s^2
J = mechanical equivalent of heat, 778 ft lb_f/Btu
P = pressure, lb_f/in.^{2}
$\dot{Q}$ = heat added, Btu/lb_m
w = velocity of the fluid, averaged over the pipe cross section, ft/s
ν = specific volume, ft^3/lb_m
W_s = shaft work, ft-lb_f/lb_m
z = elevation, ft

Pressure Losses in Straight Pipe Flow

For the flow of gases in straight pipes, the pressure drop for a given mass flowrate is complicated by the dependence of gas density on pressure. Thus, for significant pressure drops, both the velocity and the density of the gas change appreciably. Hence, to apply the Bernoulli equation properly to develop expressions to predict pressure drop, the relationship between gas pressure and density must be known. Also, the pressure drop will depend on the type of flow existing in the line. The flow will usually exist at a condition between adiabatic and isothermal.

For short insulated lines, heat transferred to or from the line is low so that the flow can be considered adiabatic. Solution of the energy balance and the Bernoulli equation for the adiabatic case assuming an ideal gas results in the following formulas:

$$\frac{4fL}{D} = \frac{1}{2k}\left[\frac{288kg_cP_1}{G^2\nu_1} + (k-1)\right]\left[1 - \left(\frac{\nu_1}{\nu_2}\right)^2\right]$$

$$+ \frac{k+1}{2k}\ln\left(\frac{\nu_1}{\nu_2}\right)^2 \tag{4-3}$$

$$\frac{P_2\nu_2}{P_1\nu_1} = \frac{T_2}{T_1} = 1 + \left[\frac{(k-1)G^2\nu_1}{288kg_cP_1}\right]\left[1 - \left(\frac{\nu_2}{\nu_1}\right)^2\right] \tag{4-4}$$

where: D = diameter of pipe, ft
$\quad\quad f$ = Fanning friction factor, dimensionless
$\quad\quad G$ = mass velocity, $lb_m/s\text{-}ft^2$
$\quad\quad k$ = ratio of specific heats, C_p/C_v
$\quad\quad L$ = length of line, ft
$\quad\quad T$ = temperature, °R

Subscripts 1, 2 refer to upstream and downstream conditions, respectively.

For long uninsulated lines, the flow will approach *isothermal* conditions. Solution of the basic equations assuming ideal gas and isothermal flow is

$$\frac{P_1^2 - P_2^2}{P_1\nu_1} = \frac{1}{144}\left[\frac{4fLG^2}{g_cD}\right]\left[1 + \frac{D}{2fL}\ln\left(\frac{P_1}{P_2}\right)\right] \tag{4-3}$$

This equation can be solved to evaluate the flow rate if the upstream and downstream pressures are known. The solution requires a trial-and-error computation if only one of these pressures is known and the pressure drop is to be calculated.

Equation 4-3 can be simplified by use of certain assumptions regarding the pressure drop. For long pipelines, the last term approaches unity (except for the case of very high pressure drops):

$$\frac{P_1^2 - P_2^2}{P_1\nu_1} = \frac{fLG^2}{36g_cD} \tag{4-4}$$

For rough estimates in cases where the pressure drop is less than 10% of the upstream pressure, this expression can be further approximated by

$$P_1 - P_2 = \frac{f\bar{\nu}LG^2}{72g_cD} \tag{4-5}$$

where $\bar{\nu}$ is the *average* specific volume of the gas, and all other terms are as previously defined.

The previous formula is based on the assumption of small pressure drop. If, however, a large pressure drop is expected, the previous expressions should be applied over a section of piping of length dL

$$\bar{\nu}dP + \frac{wdw}{g} + \lambda\frac{dL}{D}\frac{w^2}{2g} = 0 \tag{4-6}$$

where $wdw/g = d(w^2/2g)$.

The resistance coefficient λ for isothermal gas flow through a constant cross section must be constant, since it is a function of the Reynolds number ($Re = dG/\mu g$) which is also constant for a specified flow rate

$$\frac{P_1^2 - P_2^2}{2RT} = \frac{G^2}{g}\ln\frac{\bar{\nu}_2}{\bar{\nu}_1} + \frac{\lambda LG^2}{2gD} \tag{4-7}$$

Equation 4-7 permits one to determine the pressure drop $P_1 - P_2$ along a piping section of length L by trial and error. The expression simplifies to

$$P_1 - P_2 = \frac{G^2}{g\gamma} \ln \frac{\nu_2}{\nu_1} + \frac{\lambda G^2 L}{2gD\gamma} \tag{4-8}$$

This expression is also applicable to nonisothermal flows. In terms of the average specific volume rather than the average specific weight, an alternate form is

$$P_1 - P_2 = \frac{G^2}{g}(\nu_1 - \nu_2) + \frac{\lambda G^2 L}{2gD\gamma} \tag{4-9}$$

This expression can be solved for the mass throughput

$$G = \left[\frac{(P_1^2 - P_2^2)g}{2RT\left(\ln \frac{P_1}{P_2} - \frac{\lambda L}{D}\right)} \right]^{0.5} \tag{4-10}$$

The maximum mass flow rate under critical conditions can be computed from

$$G_{max} = \sqrt{\frac{P_{cr}{}^2 g}{RT}} \tag{4-11}$$

And the maximum velocity for pipe flow is

$$w_{max} = \sqrt{gP_{cr}\nu_{cr}} \tag{4-12}$$

This is the limiting gas velocity in piping which corresponds to the critical pressure at the exit. If pressure falls below this value, the gas velocity will not increase.

For frictionless adiabatic flow (i.e., the pipe system is perfectly, thermally insulated), the energy balance equation simplifies to

$$C_p(T_2 - T_1) = \frac{w_1^2}{2g}\left[\left(\frac{T_2 P_1}{T_1 P_2}\right)^2 - 1\right] \tag{4-13}$$

In practice, because of frictional resistances, $P_1 > P_2$ and consequently, a change in temperature does occur.

To develop an expression for pressure drop, the Bernoulli and Darcy-Weisbach equations are expressed in differential form, and the details are given by Cheremisinoff et al.[1]

Upon integration, we obtain

1nin

$$\lambda \frac{L}{D} = -\left(\frac{\varkappa + 1}{\varkappa}\right) \ln \frac{w_1}{w_2} + \frac{1}{\varkappa}\left(\frac{C_p^2}{w_1^2} + \frac{\varkappa - 1}{2}\right)\left(1 - \frac{w_1^2}{w_2^2}\right) \qquad (4\text{-}14)$$

where $\varkappa$ is the ratio of specific heats (C_p/C_v).

If the initial gas velocity w_1 is known, w_2 may be computed from this expression for a specified pipe length L. The value of λ does not change significantly and should only be taken as an average value when considering long lengths of piping

$$\frac{T_2}{T_1} = 1 + \frac{\varkappa - 1}{2C_p^2}w_1^2\left(1 - \frac{w_2^2}{w_1^2}\right) \qquad (4\text{-}15)$$

And from continuity,

$$\frac{P_2}{P_1} = \frac{w_1}{w_2}\left[1 + \frac{\varkappa - 1}{2C_p^2}w_1^2\left(1 - \frac{w_2^2}{w_1^2}\right)\right] \qquad (4\text{-}16)$$

Analysis of the previous equations for adiabatic gas flow in piping reveals that there is a maximum flow rate where the gas velocity at the exit reaches the velocity of sound. However, the adiabatic gas flow expressions provide essentially the same results as the isothermal analysis. For very short pipes and high pressure gradients, the adiabatic flow rate will be larger than the isothermal case; however, differences are generally no greater than 20%. Further discussions are given by Cambel et al.,[2] Shapiro,[3] and Liepmann et al.[4]

Sample Calculation 4-1. Compute the pressure drop for air flowing through a horizontal pipeline at a rate of 350 kg/hr. The pipe is 52.5mm (2 in.) in diameter, 150 m in length, and the exit pressure is atmospheric. The flow may be assumed isothermal at T = 20°C.

Solution. The cross section of the piping is

$$F = \frac{\pi \times 0.0525^2}{4} = 2.17 \times 10^{-3}m^2$$

The mass flow rate is

$$W = \frac{350}{3600} = 0.0972 \text{ kg/sec}$$

Hence, the specific weight flow rate is

$$G = \frac{W}{F} = \frac{0.0972}{2.17 \times 10^{-3}} = 44.8 \frac{kg}{m^2\text{-sec}}$$

The viscosity of air at 20°C is 0.02 cp, i.e., $\mu_g = 0.02 + 10^{-3}$kg/m-sec.

The Reynolds number is

$$Re = \frac{GD}{\mu_g} = \frac{44.8 \times 0.0525}{0.02 \times 10^{-3}} = 1.176 \times 10^5$$

The friction coefficient may be computed from the following turbulent correlation given by Perry et al.:

$$\lambda = 0.0123 + \frac{0.7544}{Re^{0.38}} = 0.0123 + \frac{0.754}{1.176 \times 10^5} = 0.021$$

The specific volume of the gas is

$$v_2 = \frac{22.4}{29} \times \frac{(273 + 20)}{273} = 0.83 \text{ m}^3/\text{kg}$$

We can now use Equation 4-9 to compute the upstream pressure:

$$P_1 = 10333 + \frac{(44.8)^2}{9.81\gamma} \ln\frac{0.83}{1} + \frac{0.022 \times (44.8)^2}{2 \times 9.81 \times 0.0525\gamma}$$

or

$$P_1 = 10333 + \frac{470}{\gamma} \ln\frac{0.83}{v_1} + \frac{6416}{\gamma}$$

This expression may be solved by a trial-and-error solution (i.e., a method of successive approximations). The gas constant R equals 1.987 kcal/mole°K, and 1 kcal equals 426.7 kg-m. Therefore,

$$R = \frac{1.987 \times 426.7}{2g} = 29.2 \frac{kg\text{-}m}{kg°K}$$

$$T - 273 + 20 - 293°K$$

$$RT = 29.2 \times 293 = 8550m$$

$$\nu_1 = \frac{RT}{P_1} = \frac{8550}{P_1}$$

The average specific weight is

$$\gamma = \frac{P_1 + P_2}{2RT} = \frac{P_1 + 10333}{17100}$$

Assuming a value for P_1 of 13,000 kg/m^2,

$$\nu_1 = \frac{8550}{13000} = 0.66 \text{ m}^3/\text{kg}$$

And the average specific weight is

$$\gamma = \frac{13000 + 10333}{17100} = 1.36 \text{ kg/m}^3$$

Substituting these values into our equation,

$$P_1 = 10333 + \frac{470}{1.36} \ln \frac{0.83}{0.66} + \frac{6416}{1.36} = 15085 \text{ kg/m}^2$$

This computed value is different from the assumed P_1 (13000) and, hence, a new P_1 should be selected and the calculations repeated. Additional computations are left to the reader.

Sample Calculation 4-2. Compute the upstream pressure for methane flowing at a rate of 1.2 kg/sec in a pipe of 130 mm ID and 30 km long. Conditions at the exit are 2.5 atm and 20°C.
Solution. For isothermal flow of a real gas, the increase in internal energy equals zero and density is constant. Hence, the equation for a differential element of piping is as follows:

$$\frac{d(w^2)}{2} + \frac{dP}{\rho} + \delta\hat{F} = 0 \tag{i}$$

where $\hat{F}$ is the friction energy per unit mass.
The change in sign on $\hat{F}$ is associated with the inverse of terms containing infinitesimal differences. Velocity may be expressed as follows:

$$w = \frac{4G}{\pi d^2 \rho}$$

or

$$d(w^2) = \left(\frac{4G}{\pi d^2}\right)2\nu d\nu$$

where ν is specific volume. Substituting this relationship into Equation (i), we obtain

$$\left(\frac{4G}{\pi d^2}\right)\frac{d\nu}{\nu} + \frac{dP}{\nu} + \frac{\lambda}{2}\left(\frac{4G}{\pi d^2}\right)^2\frac{d\ell}{d} = 0 \tag{ii}$$

To integrate (ii), the relationship between pressure, specific volume, and λ must be known.

For perfect gases at isothermal conditions

$$p\nu = \text{constant} = p_2\nu_2$$

and λ is the constant because

$$Re = \frac{Wd}{\nu} = \frac{4G}{\pi d\mu} = \text{constant}$$

Substituting this expression in (ii), we obtain the following:

$$\left(\frac{4G}{\pi d^2}\right)^2\frac{dp}{p} + \frac{pdp}{p_2\nu_2} + \frac{\lambda}{2}\left(\frac{4G}{\pi d^2}\right)^2\frac{d\ell}{d} = 0$$

Integrating over the limits of 0 to ℓ and between P_1 and P_2, we obtain

$$\left(\frac{4G}{\pi d^2}\right)^2\left(\ln\frac{P_1}{P_2} + \frac{\lambda}{2}\frac{\ell}{d}\right) = \frac{P_1^2 - P_2^2}{P_2\nu_2} \tag{iii}$$

To evaluate the friction factor, first compute the Reynolds number

$$Re = \frac{4G}{\pi d\mu} = \frac{4 \times 1.2}{\pi \times 0.13 \times 1.08 \times 10^{-5}} = 1.09 \times 10^6$$

Using a roughness coefficient e = 0.1 mm and relative roughness e/d = 0.1/130 = 0.00077, from the Moody plot (see Figure 3-1) the value of λ is equal to 0.018.

The specific volume of methane at 2.5 atm, 20°C is

$$v_2 = \frac{22.4}{16} \times \frac{1.033}{2.5} \times \frac{273 + 20}{273} = 0.62 \text{ m}^3/\text{kg}$$

Substituting these values into Equation (iii),

$$\left(\frac{4 \times 1.2}{\pi \times 0.13^2}\right)^2 \left(\ln \frac{P_1}{2.5 \times 9.81 \times 10^4} + \frac{0.018 \times 30{,}000}{2 \times 0.13}\right)$$

$$= \frac{P_1^2 - (2.5 \times 9.81 \times 10^4)^2}{2.5 \times 9.81 \times 10^4 \times 0.62}$$

To simplify calculations, neglect the term $\ln P_1/P_2$, assuming it to be small in comparison to $\lambda/2 \times \ell/d$.

Hence,

$$P_1 = \left\{(2.5 \times 9.81 \times 10^4)^2 + 2.5 \times 9.81 \times 10^4 \right.$$

$$\left. \times 0.62\left(\frac{4 \times 1.2}{\pi \times 0.13^2}\right)^2 \left(\frac{0.018 \times 30000}{2 \times 0.13}\right)\right\}^{1/2}$$

$$= 1.625 \times 10^6 \text{N/m}^2 = 16.6 \text{ atm}$$

Checking the error obtained as a result of this assumption,

$$\ln \frac{P_1}{P_2} = 1.89 << \frac{\lambda}{2} \times \frac{\ell}{d} = \frac{0.018 \times 30000}{2 \times 0.13} = 2075$$

Hence, the calculation based on the initial pressure is sufficiently accurate.

Discharge of Gases

The discharge of gases through orifices or nozzles may be approximated as frictionless adiabatic flow. The reasons for this are: (1) friction losses are minor because of the short distances traveled, and (2) heat transfer is negligible ($\dot{Q} = 0$) because the changes that the gas undergoes are slow enough to keep velocity and temperature gradients small.

Applying this approximation, simplification of the energy balance expression can be made for these flow systems. For the frequently encountered system of gas discharge from a tank, the following formula can be used to estimate the velocity at the point of discharge:

$$w_2 = \sqrt{2g(i_1 - i_2)} \tag{4-17}$$

Where i represents the gas enthalpy or noting that $i_1 - i_2 = C_p(T_1 - T_2)$,

$$w_2 = \sqrt{C_p(T_1 - T_2)2g} \tag{4-18}$$

From the ideal gas law ($P_1\nu_1 = RT_1$) and the thermodynamic relationship $C_p - C_v = R$, an expression for the velocity of reversible adiabatic discharge of an ideal gas is

$$w_2 = \sqrt{2g\frac{\varkappa}{\varkappa - 1}(P_1\nu_1)\left[1 - \left(\frac{P_2}{P_1}\right)^{\frac{\varkappa - 1}{\varkappa}}\right]} \tag{4-19}$$

The mass gas rate per unit discharge area may be expressed in terms of the discharge velocity

$$G = w_2\gamma_2 = \frac{w_2}{\nu_2} \tag{4-20}$$

where γ_2 is the specific weight of gas (i.e., the inverse of specific volume).

Combining Equations 4-19 and 4-20 and including the expression for a reversible adiabatic flow ($P_1\nu_1{}^\varkappa = P_2\nu_2{}^\varkappa$), the following equation defines the mass rate in terms of the thermodynamic properties of the fluid:

$$G = \sqrt{2g\frac{\varkappa}{\varkappa - 1}\frac{P_1}{\nu_1}\left[\left(\frac{P_2}{P_1}\right)^{2/\varkappa} - \left(\frac{P_2}{P_1}\right)^{\frac{\varkappa - 1}{\varkappa}}\right]} \tag{4-21}$$

Equation 4-21 has a maximum with respect to P_2 or P_2/P_1. The pressure at which the maximum flow rate occurs (i.e., the "critical" pressure) is

$$P_{cr} = P_1\left(\frac{2}{\varkappa + 1}\right)^{\frac{\varkappa}{\varkappa - 1}} \qquad (4\text{-}22)$$

Because $\varkappa$ for gases does not change appreciably, it may be assumed that the critical pressure is in the range of 0.53 P_1 to 0.58 P_1 (i.e., the critical pressure is approximately one-half of the tank pressure).

Knowledge of the critical pressure is important for evaluating the efficiency of the flow process. If, for example, the pressure at the exit is higher than the critical value computed from Equation 4-22, then the flow rate in the orifice will not reach its maximum value. At complete expansion, the gas velocity may be computed from Equation 4-19. If the exit pressure of the orifice is less than the critical value (because the maximum flow rate was exceeded), the amount of discharge must reach a maximum value; that is, the critical pressure will be achieved. Thus, further gas expansion will occur downstream of the orifice. The flow will expand and consequently its head will decrease. Regardless, at $P_2 < P_{cr}$, the gas velocity will not correspond to pressure P_2 and Equation 4-19 should not be used. The discharge velocity will reach a lower value corresponding to P_{cr}. This critical velocity is determined by replacing P_2 in Equation 4-19 with P_{cr}.

The critical discharge velocity corresponding to the critical pressure is

$$W_{cr} = \sqrt{g\varkappa P_{cr}\nu_{cr}} \qquad (4\text{-}23)$$

Equation 4-23 is the expression for the "sound velocity." Therefore, the maximum linear discharge velocity from a tank orifice is equal to the sound velocity which is related to a corresponding temperature and pressure at the exit. This temperature is

$$T_{cr} = T_1\left(\frac{2}{\varkappa + 1}\right) \qquad (4\text{-}24)$$

At any point in the expanding flow the weight velocity is constant and flow rate changes according to Equation 4-21. Thus, the cross section of the gas flow (W/G) will also vary.

It is possible to design a nozzle with variable cross section for a specified weight flow rate using Equation 4-21. The critical pressure and velocity will be attained at the narrowest section of the nozzle. In the expanding section pressure decreases and velocity increases and, according to Equation 4-19, the flow eventually reaches supersonic conditions. The use of a nozzle of variable cross section is logical when the counterpressure (downstream) is less than critical pressure P_{cr}. If the downstream pressure exceeds P_{cr}, the effect is still the same since the orifice will have a diameter equal to the largest size of the nozzle's diameter.

The previous discussion assumes thermodynamically reversible gas discharge through an orifice or nozzle (i.e., neglecting friction). However, friction can play an important role during discharge in some applications. Friction associated with the kinetic energy of the gas is converted into heat, thus increasing the enthalpy of the discharge. Enthalpy will, therefore, be less than the case of reversible flow at the same pressure and counterpressure. Futhermore, Equation 4-17 shows that the discharge gas velocity will decrease. To compute the discharge velocity for this case (without considering counterpressure), the temperature of the discharging gas is needed. Knowing the exit temperature T_2 and counterpressure P_2, we find the enthalpy i_2 and, from Equation 4-17, the discharge gas velocity is obtained. The following sample calculation illustrates the use of these equations.

Sample Calculation 4-3. A reaction takes place in a nitrogen atmosphere in an autoclave at t = 130°C and p = 10 atm. The volume of the gas is 26.4 gallons. After 10 minutes, it was discovered that the pressure dropped to 9.7 atm because of leakage. The autoclave was found to be rated for pressure only to 6 atm. Estimate how long it will take the vessel's pressure to drop to 6 atm and the rate at which nitrogen is escaping.

Solution. Assume the gas discharge occurs through a narrow slit and is adiabatic and reversible. First, determine the pressure in the autoclave. For nitrogen, C_v = 5 cal/mole and C_p = C_v + R = 7 cal/mole. Hence,

$$x = C_p/C_v = 7/5 = 1.4$$

The critical pressure is

$$P_{cr} = P_1 \left(\frac{2}{x + 1} \right)^{\frac{x}{x - 1}} = P_1 \left(\frac{2}{1.4 + 1} \right)^{\frac{1.4}{1.4 - 1}} = 0.53 \ P_1$$

Note that the critical pressure will always be greater than the surroundings to which the nitrogen is expanding. The gas discharge will occur at the sound velocity.

The absolute temperature in the autoclave is $T_1 = 273 + 130 = 430°K$; and from Equation 4-24,

$$T_{cr} = 403\left(\frac{2}{1.4 + 1}\right) = 337°K$$

The specific volume of nitrogen at T_{cr} and P_{cr} is as follows (molecular weight of nitrogen is 28, and the volume of one kg-mole is 22.4 m³):

$$\nu_{cr} = \frac{22.4}{28} \times \frac{337}{273} \times \frac{10^4}{0.53\ P_1} = \frac{18650}{P_1}\ m^3/kg$$

where 10^4 kg/m² is the atmospheric pressure. Pressure P_1 is expressed in units of kg/m². As noted, the discharge velocity through the slit may be computed from Equation 4-23:

$$w_{cr} = \sqrt{9.81 \times 1.4 \times 0.53\ P_1 \times \frac{18650}{P_1}}$$

$$= 368\ m/sec\ (1207\ fps)$$

The nitrogen weight rate at discharge is

$$G = \frac{w_{cr}}{\nu_{cr}} = \frac{368}{18650}P_1 = 0.0197\ P_1\ \frac{kg}{m^2\text{-sec}}$$

When the pressure is P_1, the specific volume of the nitrogen is

$$\nu = \frac{22.4}{28} \times \frac{403}{273} \times \frac{10^4}{P} = \frac{11850}{P}$$

Because the volume of the autoclave is 26.4 gallons (0.1 m³), the weight of nitrogen is

$$W = \frac{0.1\ P_1}{11850} = 8.42 \times 10^{-6}P$$

The nitrogen loss accompanied by the decreased dP is

$$dW = -8.42 \times 10^{-6}dP$$

The leakage occurs when the pressure decreases by some amount of dP over a time interval dt through a slit having cross-sectional area F. The discharge can be expressed by a rate expression:

$$dW = Gdt$$

or

$$-8.42 \times 10^{-6}dP = 0.0197PFdt$$

Integrating this expression over the pressure limits of 10 atm to P_1, the time over which the pressure drop occurs is

$$t = \frac{-4.27 \times 10^{-4}}{F} \int_{10}^{P_1} \frac{dP}{dt} = 9.48\frac{10^{-4}}{F} \log \frac{10}{P_1}$$

where P is expressed in units of absolute atmospheres and t in seconds.

For $t = 10$ min $= 600$ sec, the pressure in the autoclave is 9.7 atm. Hence, the area of discharge in mm^2 is

$$F = \left(\frac{9.84 \times 10^{-4}}{600} \log \frac{10}{9.7}\right)10^6 = 0.0213 \text{ mm}^2$$

The time required for the pressure to drop to 6 atm is

$$t = \frac{9.84 \times 10^{-4}}{0.0213 \times 10^{-6}} \log \frac{10}{6} = 10250 \text{ sec. } (\sim 3 \text{ hrs})$$

Miscellaneous Data

Additional data on pressure losses are given in this section. Table 4-1 provides pressure drop data for air flows through straight (Schedule 40) pipe. For pipe lengths other than the 100-foot reference given in Table 4-1, the pressure drop is proportional to the length. For example, for 50 feet of pipe, the air pressure drop is approximately one-half of the value in the table. Pressure drop is inversely proportional to the absolute pressure and directly proportional to the absolute temperature.

To estimate pressure drops for inlet or average air pressures and temperatures other than 100 psi and 60°F, table values should be corrected for by the following ratio:

Table 4-1
Values of Air Pressure Drop in Straight Pipe

at 60°F and 14.7 psia	Compressed Air at 60°F and 100 psig	Air Pressure Drop In Pounds per Square Inch Per 100 Feet of Schedule 40 Pipe For Air at 100 PSIG and 60°F									
		1/4"	3/8"	1/2"	3/4"	1"	1-1/4"	1-1/2"	2"	2-1/2"	3"
1	0.128	0.083	0.018								
2	0.256	0.285	0.064	0.020							
3	0.384	0.605	0.133	0.042							
4	0.513	1.04	0.226	0.071							
5	0.641	1.58	0.343	0.106	0.027						
6	0.769	2.23	0.408	0.148	0.037						
8	1.025	3.89	0.848	0.255	0.062	0.019					
10	1.282	5.96	1.26	0.356	0.094	0.029					
15	1.922	13.0	2.73	0.334	0.201	0.062					
20	2.563	22.8	4.76	1.43	0.345	0.102	0.026				
25	3.204		7.34	2.21	0.526	0.156	0.039	0.019			
30	3.845		10.5	3.15	0.748	0.219	0.055	0.026			
35	4.486		14.2	4.24	1.00	0.293	0.073	0.035			
40	5.126		18.4	5.49	1.30	0.379	0.095	0.044			
45	5.767		23.1	6.90	1.62	0.474	0.116	0.055			
50	6.408			8.49	1.99	0.578	0.149	0.067	0.019		
60	7.690			12.2	2.85	0.819	0.200	0.094	0.027		
70	8.971			16.5	3.83	1.10	0.270	0.126	0.036		
80	10.25			21.4	4.96	1.43	0.350	0.162	0.046	0.019	
90	11.53			27.0	6.25	1.80	0.437	0.203	0.058	0.023	
100	12.82				7.69	2.21	0.534	0.247	0.070	0.029	
125	16.02				11.9	3.39	0.825	0.380	0.107	0.044	
150	19.22				17.0	4.87	1.17	0.537	0.151	0.062	0.021
175	22.43				23.1	6.60	1.58	0.727	0.205	0.083	0.028
200	25.63				30.0	8.54	2.05	0.937	0.264	0.107	0.036
225	28.84					10.8	2.59	1.19	0.331	0.134	0.045
250	32.04					13.3	3.18	1.45	0.404	0.164	0.055
275	35.24					16.0	3.83	1.75	0.484	0.191	0.066
300	38.45					19.0	4.56	2.07	0.573	0.232	0.078
325	41.65					22.3	5.32	2.42	0.673	0.270	0.090
350	44.87					25.8	6.17	2.80	0.776	0.313	0.104
375	48.06					29.6	7.05	3.20	0.887	0.356	0.119
400	51.26					33.6	8.02	3.64	1.00	0.402	0.134
425	54.47					37.9	9.01	4.09	1.13	0.452	0.151
450	57.67						10.2	4.59	1.26	0.507	0.168
475	60.88						11.3	5.09	1.40	0.562	0.187
500	64.08						12.5	5.61	1.55	0.623	0.206
550	70.49						15.1	6.79	1.87	0.749	0.248
600	76.90						18.0	8.04	2.21	0.887	0.293
650	83.30						21.1	9.43	2.60	1.04	0.342
700	89.71						24.3	10.9	3.00	1.19	0.395
750	96.12						27.9	12.6	3.44	1.36	0.451
800	102.5						31.8	14.2	3.90	1.55	0.513
850	108.9						35.9	16.0	4.40	1.74	0.576
900	115.3						40.2	18.0	4.91	1.95	0.642
950	121.8							20.0	5.47	2.18	0.715
1,000	128.2							22.1	6.06	2.40	0.788
1,100	141.0							26.7	7.29	2.89	0.948
1,200	153.8							31.8	8.63	3.44	1.13
1,300	166.6							37.3	10.1	4.01	1.32
1,400	179.4								11.8	4.65	1.52
1,500	192.2								13.5	5.31	1.74
1,600	205.1								15.3	6.04	1.97
1,800	203.7								19.3	7.65	2.50
2,000	256.3								23.9	9.44	3.06

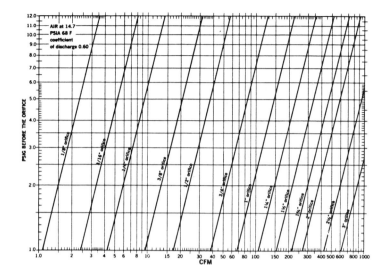

Figure 4-1. Chart for estimating air discharge through an orifice in cfm (free air at standard atmospheric pressure).

$$\left(\frac{100 + 14.7}{P + 14.7}\right) \text{ and } \left(\frac{460 + t}{520}\right)$$

where: P = inlet or average gauge pressure, psi
        t = temperature, °F

Compressed air flows (in cfm) at any temperature and pressure other than STP must be corrected by the value of cfm of free air by the ratio

$$\left(\frac{14.7}{14.7 + P}\right) \text{ and } \left(\frac{460 + t}{520}\right)$$

Figure 4-1 provides a chart for estimating air discharge through an orifice (discharge coefficient = 0.6) for standard conditions. Finally, Table 4-2 provides physical properties data for selected gases.

Table 4-2
Physical Properties of Different Gases

NAME OF GAS	OXYGEN	NITROGEN	ARGON.	HELIUM
Chemical Symbol	O_2	N_2	Ar	He
Molecular Weight	31.9988	28.0134	39.948	4.0026
Color	None	None	None	None
Odor	None	None	None	None
Taste	None	None	None	None
Spec. Gravity (Air=1) 70°F. 1 Atm.	1.105	0.9669	1.380	0.13796
Density. Lb. per Cu. Ft. 70°F.1 Atm.	0.08281	0.07245	0.1034	0.01034
Spec. Vol. Cu. Ft. per Lb. 70°F. 1 Atm.	12.076	13.803	9.671	96.71
Density Sat'd Vapor, Lb. per Cu. Ft. 1 Atm.	0.27876	0.2874	0.35976	1.0434
Normal Boiling Point °F	−297.33	−320.36	−302.55	−452.1
Heat of Vaporization BTU per Pound	91.7	85.6	70.1	9.0
Critical Pressure Atmospheres, Abs. Lb. per Sq. In., Abs.	50.14 736.9	33.54 492.9	48.34 710.4	2.26 33.2
Critical Temp. °F.	−181.08	−232.40	−188.12	−450.31
Triple Point Pressure Atmosphere, Abs. Lb. per Sq. In., Abs.	0.00145 0.0213	0.1238 1.819	0.68005 9.9940	None
Triple Point Temp. °F.	−361.83	−346.01	−308.8	None
Specific Heat. Const. Press	0.2199 @77°F	0.2488 @77°F	0.1244 @77°F	1.2404 @77°F
Ratio Specific Heats	1.396 @80.3°F	1.4014 @70°F	1.6665 @86°F	1.6671 @77°F
Coeff. Viscosity, Micropoises @77°F.	206.39	177.96	226.38	198.5
Thermal Cunductivity, 32°F. BTU/(Sq. Ft.) (Hr.) (°F/Ft.)	0.0142	0.0139	0.00980	0.08266 @40°F
Ionization Potential, Volts	13.6	14.5	15.7	24.5
Excitation Potentials: First Resonance Potential, Volts	9.1	6.3	11.56	20.91
Metastable Potentials, Volts			11.66 11.49	19.77

* Normal Sublimation Temperature.
** Latent Heat of Sublimation.

(Table 4-2 continued)

ACETYLENE C_2H_2 26.0382	HYDROGEN H_2 2.01594	NEON Ne 20.183	KRYPTON Kr 83.80	XENON Xe 131.30	AIR – 28.96
None	None	None	None	None	None
Sweet	None	None	None	None	None
None	None	None	None	None	None
0.9053	0.0695	0.6958	2.898	4.56	1.0
0.06785	0.005209	0.05215	0.2172	0.3416	0.07493
14.7	192.0	19.175	4.604	2.927	13.3
0.10800	0.083133	0.5963	0.536	0.718	
−118.5°	−423.0	−410.8	−243.8	−162.5	
344.8°°	191.7	38.3	46.3	41.4	88.3
60.58	12.98	26.19	54.3	57.64	
890.3	190.8	384.9	798.	847.1	
+95.32	−399.96	−379.75	−82.8	+61.86	
1.2651	0.071	0.4273	0.7220	0.8064	
18.592	1.04	6.28	10.61	11.85	
−112.99	−434.56	−415.49	−251.28	−169.18	
0.4067 @80°F	3.4202 @77°F	0.2462 @77°F	0.0597 @87°F	0.0382 @77°F	0.2406 @80.3°F
1.234 @77°F	1.405 @77°F	1.642 @68°F	1.701 @87°F	1.666 @68°F	1.4017 @80.3°F
95.5	89.37	313.81@68°F	251.71	231.02	184.67@80°F
0.0123 @80°F	0.0973	0.021087	0.00501	0.00293	0.0139
11.6	13.5	21.5	13.9	12.1	
	10.2	16.58	9.98	8.39	
		16.62	10.51	9.4	
		16.53	9.86	8.28	

References

1. Cheremisinoff, N. P. and Azbel, D., *Fluid Mechanics and Unit Operations,* Ann Arbor Science Pub., Ann Arbor, MI (1983).
2. Cambel, A. B., and Jennings, B. J., *Gas Dynamics,* McGraw-Hill Book Co., NY (1958).
3. Shapiro, A. H., *The Dynamics and Thermodynamics of Compressible Fluid Flow,* Roland Pub. Co., New York, (1953).
4. Liepmann, H. W. and Roshko, A., *Elements of Gas Dynamics,* John Wiley & Sons, Inc., New York (1957).

5

DISCHARGE
THROUGH VARIABLE
HEAD METERS

Types of Variable Head Meters

Evaluation of head losses is an important practical problem connected with the calculation of the energy required for fluid displacement in pumps, compressors, etc., as well as for measuring flow quantities entering and leaving process equipment. Many flow problems in this category can be addressed through the Bernoulli equation. Some practical applications of the Bernoulli equation as applied to the use of variable-head meters are commented on in this section. Such devices are extensively used throughout industry to measure and control flows through equipment.

Notes are limited to the three most widely used headmeters, namely, pitot tubes, orifice, and Venturi meters. Extensive data on the design of these devices are given by the American Society of Mechanical Engineers (ASME Publications[1]) and applications to industrial problems are given by Cheremisinoff.[2,3]

The *pitot tube* is used for measuring local fluid velocities. The device consists of a stainless steel tube with its inlet opening turned upstream into the flow. The inlet, therefore, receives the full impact of the flow against it. The impact is completely converted into pressure head $w^2/2g$ and superimposed on the existing static pressure of the fluid. In principle, the pitot tube consists of both an *impact tube* and a *piezometer tube* (refer to Figure 5-1a). Because an impact tube is used in connection with a piezometer tube, the static pressure may be subtracted from the total pressure measured and the difference is the velocity head. This is illustrated in Figure 5-2. The pressure difference is conveniently measured with a differential U-tube as shown in Figure 5-1a. Note that the U-tube should contain a liquid which does not mix with the working

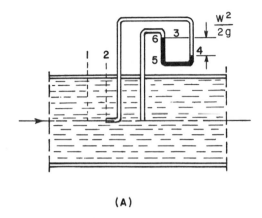

(A)

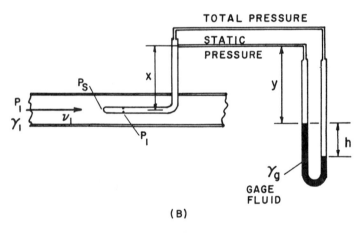

(B)

Figure 5-1. Pitot tube used to measure fluid velocity in a pipe.

fluid and has higher density. Both the impact and piezometer tube are combined into a standard S-shaped pitot tube by including static pressure taps downstream of the impact tube's tip. The design is illustrated in Figure 5-1b.

An *orifice meter* consists of a thin plate mounted between two flanges with an accurately drilled hole positioned concentric with the pipe axis. The principle of measurement for the orifice meter is based on the re-

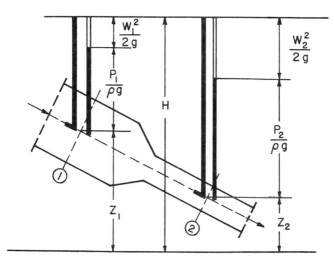

Figure 5-2. Ideal fluid flowing through a variable cross section arbitrarily located in space.

duction of flow pressure accompanied by an increase in velocity (i.e., Bernoulli's principle). The reduction of the cross section of the flowing stream as it passes through the orifice increases the velocity head at the expense of pressure head. Figure 5-3 illustrates the operation of an orifice meter whereby a *manometer* or pressure gauges are used to measure the upstream and downstream pressures. By applying the Bernoulli equation, the discharge can be determined from the manometer's reading based on the known area of the orifice.

A *Venturi meter,* illustrated in Figure 5-4, consists of a short length of straight tubing connected at either end of the pipe by conical sections. The measurement principle is based on the reduction of flow pressure accompanied by an increase in velocity of the Venturi throat. The pressure drop experienced in the upstream cone section is used to measure the rate of flow through the Venturi meter. On the discharge side of the meter, the fluid velocity is decreased and the original pressure is recovered. Because of its shape, pressure losses in a Venturi meter are less than in an orifice meter. However, a Venturi is large in comparison to an orifice meter which can be readily mounted between flanges.

Because of its size and expense, a smaller version of the Venturi has been developed called the *flow nozzle.* A typical flow nozzle is illus-

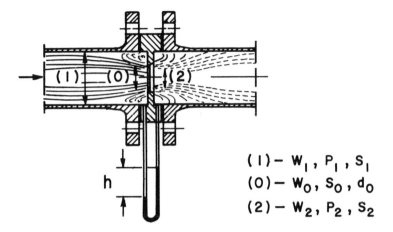

$$(1) - W_1, P_1, S_1$$
$$(0) - W_0, S_0, d_0$$
$$(2) - W_2, P_2, S_2$$

Figure 5-3. An orifice meter in operation.

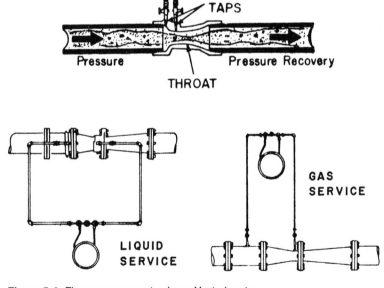

Figure 5-4. Flow measurement using a Venturi meter.

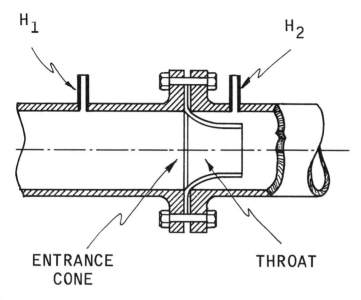

Figure 5-5. A typical flow nozzle.

trated in Figure 5-5. It is often used as the primary element for measuring liquid flows. In the design, the diverging exit of the Venturi meter is omitted and the converging entrance altered to a more rounded configuration. In both the Venturi meter and flow nozzle, the cross-sectional area of the compressed stream ($S_2 = \pi d_1^2/4$ is equal to the cross-sectional area of the hole $S_0 = \pi d^2/4$. Formulas applicable to orifice meters, Venturi meters, and flow nozzles are outlined next.

Basis of Head Loss Calculations

A special form of the Bernoulli theorem for any two cross-sections of a flow element is

$$z_1 + \frac{P_1}{\rho g} + \frac{w_1^2}{2\alpha g} = z_2 + \frac{P_2}{\rho g} + \frac{w_2^2}{2\alpha g} \tag{5-1}$$

Equation 5-1 is related not to the cross section as a whole, but to any pair of compatible points in these sections (for example, points located along the axis of the piping). To compare congruent values over total cross sections, it must be assumed that the terms in the equation may be approximated, which represents the point form of the Bernoulli equation for ideal fluids. It follows that the *hydrodynamic head* ($z + P/g + w^2/2g$) is constant for all cross sections of ideal steady flow. Z is the *leveling height* (also called the *geometric head*) which represents a specific potential energy at a given point within the cross section. $P/\rho g$ is the static or piezometric head characterizing the specific potential energy of pressure at a given point. Both terms may be expressed either in units of length or in units of specific energy (i.e., energy per unit weight of fluid).

The third item, $w^2/2g$, is also expressed in units of length:

$$\left(\frac{w^2}{2g}\right) \equiv \left(\frac{m^2\text{-sec}^2}{\text{sec}^2\text{-m}}\right) \equiv (m)$$

or, multiplying and dividing by unit weight, in units of energy $w^2/2g$ is called the *velocity or dynamic head* which characterizes a specific kinetic energy at a given point within the cross section.

According to the Bernoulli equation, the sum of the static and dynamic heads and level does not change from one cross section of flow to another. This statement is inclusive only of steady-state flow of an ideal fluid. It further follows that the sum of the potential and kinetic energies is constant.

A conversion of energy occurs when the flow changes. This change in energy is reflected in the fluid velocity. Reducing a pipe's cross-sectional area causes part of the potential energy of pressure to be converted into kinetic energy, and vice versa. When the flow area is increased, part of the kinetic energy is converted into potential; however, the total energy is unchanged. Hence, we may conclude that the amount of energy entering the initial cross section is equal to that leaving the pipe.

Refer back to Figure 5-2. Points on the flow axis at sections 1 and 2 are at heights z_1 and z_2 above the datum, respectively. At each of these points two piezometric tubes are inserted into the flow. One tube at each point has its end into the direction of flow. The levels of fluid in the straight, vertical tubes settle at heights corresponding to the hydrostatic pressures at the points of their submergence (i.e., they provide a measure of the *static head* at corresponding points). The fluid height in the bent tubes is higher than in the straight ones since the measurement represents the sum of the static and dynamic heads. The levels in the bent tubes are the same since they are both referenced to the same datum plane and are a measure of the hydrodynamic head.

Since the flow cross section at plane 2-2 is less than that of 1-1, from the continuity, fluid velocity w_2 (for a constant flow rate) must be greater than w_1. Hence, the kinetic energy at 2-2 is greater than 1-1 ($w_2^2/2g > w_1^2/2g$). Therefore, the difference between static and dynamic heads at plane 2-2 is greater than the difference at plane 1-1.

The Bernoulli equation states that the fluid level in the straight tube at 2-2 is less than the corresponding height in the straight tube in plane 1-1 and, by the same value, the velocity head at 2-2 exceeds 1-1. This illustrates the mutual conversion of potential energy into kinetic when the flow cross section is changed. The overall conclusion is that the sum of these energies in any cross section of the piping remains unchanged.

For real fluids, both shear and friction forces are important. Friction forces exert resistance to fluid motion. A portion of the flow energy must be devoted to overcome this *hydraulic resistance*. The total energy decreases continuously in the direction of flow as a portion is converted from potential energy into *lost energy* (the energy expended to overcome friction). This conversion is irreversible and is lost in the form of heat dissipation to the surroundings. For the system just analyzed, this means that

$$z_1 + \frac{P_1}{\rho g} + \frac{w_1^2}{2g} > z_2 + \frac{P_2}{\rho g} + \frac{w_2^2}{2g}$$

For real (viscous) fluids, the levels in the bent tubes at planes 1-1 and 2-2 in Figure 5-2 are not the same. This difference in levels is attributed to energy losses in the fluid path from 1-1 to 2-2 and is referred to as the *lost head* h_ℓ. Hence, the Bernoulli equation is corrected to

$$z_1 + \frac{P_1}{\rho g} + \frac{w_1^2}{2g} = z_2 + \frac{P_2}{\rho g} + \frac{w_2^2}{2g} + h_\ell \qquad (5\text{-}2)$$

where the lost head h_ℓ characterizes the specific energy spent for overcoming hydraulic resistances.

Another form of this expression can be obtained by multiplying both sides by ρg:

$$\rho g z_1 + \frac{\rho w_1^2}{2\alpha} + P_1 = \rho g z_2 + P_2 + \frac{\rho w_2^2}{2\alpha} + \Delta P_\ell \qquad (5\text{-}3)$$

ΔP_ℓ is the lost pressure drop

$$\Delta P_\ell = \rho g h_\ell \tag{5-4}$$

Head Meter Formulas

For a pitot tube, the maximum flow velocity along a pipe axis may be computed from the measured pressure head $w^2/2g$ using the Bernoulli equation. Consider an incompressible fluid flowing between points 1 and 2 in Figure 5-1a. Since the velocity at point 2 is zero,

$$\frac{w_1^2}{2g} = \frac{P_2 - P_1}{\rho} \tag{5-5}$$

or

$$w = C\sqrt{\frac{2g\Delta P}{\rho}} \tag{5-6}$$

where C is the pitot coefficient (obtained from calibration). For a well-designed pitot tube, C has a value between 0.96 and 0.98.

The value of ΔP may be obtained from the Bernoulli equation for the case of zero flow:

$$\Delta P = -\rho \Delta z g \tag{5-7}$$

The final form of this expression (see Cheremisinoff[4] for derivation) is

$$w = C\sqrt{\frac{2g(\rho_m - \rho)\Delta h}{\rho}} \tag{5-7a}$$

where ρ_m is the density of the manometer liquid and $h = z_3 - z_4$.

To evaluate the mean fluid velocity in a pipe from local measurements, follow this procedure:

Step 1. Divide the pipe's cross section into a number of equal areas.
Step 2. Measure the local velocity at a representative point in each small area.
Step 3. Determine the mean value by averaging results.

For rectangular ducts, the cross section can be divided into small squares or rectangles and the local velocity measured at the center of each.

For an N-point traverse on a circular cross section, measurements should be taken on each side of the cross section at:

$$100 \times \sqrt{\frac{2n - 1}{N}}, \% (n = 1, 2, 3, \ldots N/2) \qquad (5\text{-}8)$$

For a normal velocity distribution in a circular pipe, a 10-point traverse gives a mean velocity about 0.3% overpredicted. A 20-point traverse gives a 0.1% overestimate. Figures 5-6 and 5-7 show recommended traversing positions for circular and rectangular cross sections, respectively.

For an orifice, the following equation can be used to estimate downstream velocity:

$$w_L = \sqrt{\frac{2g\Delta h}{1 - (d_2/d_1)^4}} \qquad (5\text{-}9)$$

where

$$\Delta h = \frac{w_2^2}{2g} - \frac{w_2^2}{2g}\left(\frac{d_2}{d_1}\right)^4 \qquad (5\text{-}10)$$

The volume liquid rate Q in section S_o (area) of the hole in the orifice meter (and consequently, in the piping) is

$$Q = \frac{C\pi}{4}d_0^2\sqrt{2g\Delta h/\left(1 - \left(\frac{d_0}{d_1}\right)^4\right)} \qquad (5\text{-}11)$$

where C is the discharge coefficient and a function of the Reynolds number. Figure 5-8 gives typical discharge coefficients for different orifices.

The term (d_2/d_1) in the denominator of Equation 5-11 is usually small. Hence, as a first approximation, the volumetric flow rate may be computed from

$$Q = \frac{\alpha'\pi}{4}d_0^2\sqrt{2g\Delta h} \qquad (5\text{-}12)$$

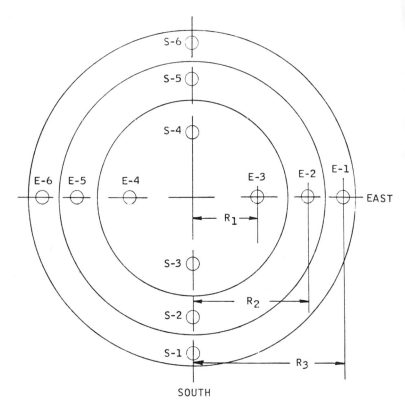

Figure 5-6. Traversing in circular pipes.

Also, the mean velocity through the pipe may be approximated by

$$w = \alpha' \left(\frac{d_0}{d}\right)^2 \sqrt{2g\Delta h} \tag{5-13}$$

Table 5-1 gives flow capacities of water through circular, straight-edge orifices.

Rotameters

The rotameter is a variable-area meter, as shown in Figure 5-9. It consists of a tapered vertical tube in which fluid enters through the bot-

(Text continued on page 92)

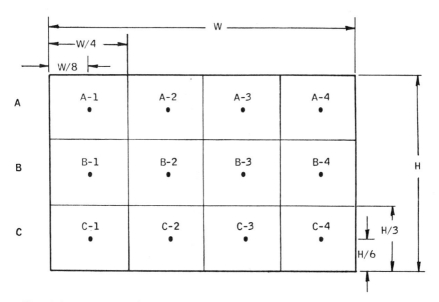

The minimum number of test points for uniform flow may be determined as follows:

Cross-Sectional Area (ft²)	Number Of Test Points
<2	4
2–25	12
>25	20

Figure 5-7. Traversing in rectangular ducts.

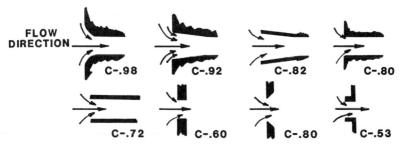

Figure 5-8. Orifice discharge coefficients.

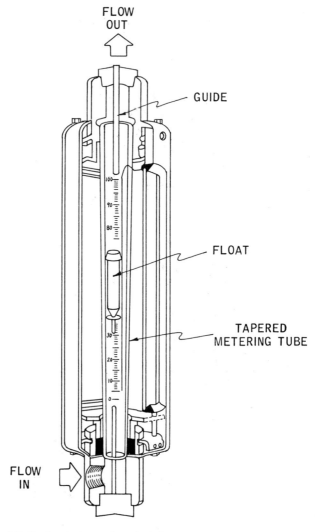

Figure 5-9. Design features of a rotameter.

Table 5-1
Capacity Data on Water Flow Through Circular, Straight-Edge Orifices[5]

			Diameter of			
	Water	Velocity Through Orifice	$1/64$	$1/32$	$1/16$	$1/8$
psi	(ft)	(fps)		Flow Through		
1	2.3	7.44	0.0044	0.0178	0.0711	0.285
2	4.6	10.52	0.0063	0.0252	0.101	0.404
3	6.9	12.89	0.0077	0.0309	0.123	0.494
4	9.2	14.88	0.0089	0.0357	0.142	0.570
5	11.6	16.64	0.0099	0.0398	0.159	0.637
6	13.9	18.22	0.0108	0.0437	0.174	0.699
7	16.2	19.69	0.0117	0.0472	0.188	0.754
8	18.5	21.04	0.0125	0.0505	0.201	0.806
9	20.8	22.32	0.0133	0.0535	0.213	0.855
10	23.1	23.53	0.0140	0.0564	0.224	0.900
12	27.7	25.77	0.0153	0.0618	0.246	0.988
14	32.3	27.84	0.0165	0.0668	0.266	1.07
16	37.03	29.76	0.0177	0.0714	0.284	1.14
18	41.6	31.57	0.0188	0.0757	0.301	1.21
20	46.2	33.27	0.0198	0.0798	0.318	1.27
25	57.7	37.20	0.0221	0.0892	0.355	1.42
30	69.3	40.75	0.0242	0.0978	0.390	1.56
35	80.8	44.02	0.0261	0.105	0.420	1.69
40	92.4	47.06	0.0279	0.113	0.450	1.80
45	104	49.91	0.0296	0.120	0.476	1.91
50	116	52.61	0.0312	0.126	0.503	2.02
60	139	57.63	0.0343	0.138	0.550	2.20
70	162	62.25	0.0370	0.149	0.594	2.38
80	185	66.55	0.0395	0.159	0.635	2.55
90	208	70.58	0.0419	0.169	0.674	2.70
100	231	74.40	0.0442	0.178	0.712	2.85
125	289	83.18	0.0494	0.199	0.794	3.19
150	346	91.13	0.0542	0.218	0.871	3.49
175	404	98.42	0.0586	0.236	0.940	3.77
200	462	105.2	0.0626	0.252	1.01	4.04
225	520	111.6	0.0664	0.268	1.08	4.28
250	577	117.6	0.0700	0.282	1.13	4.50
300	693	128.9	0.0767	0.309	1.23	4.94
350	808	139.2	0.0828	0.334	1.33	5.32
400	924	148.8	0.0886	0.357	1.42	5.70
450	1039	157.8	0.0940	0.379	1.51	6.05
500	1155	166.4	0.0990	0.399	1.59	6.37
600	1386	182.2	0.108	0.437	1.74	6.99
700	1617	196.9	0.117	0.472	1.88	7.54
800	1848	210.4	0.125	0.505	2.01	8.06
900	2079	223.2	0.133	0.535	2.13	8.55
1000	2310	235.3	0.140	0.564	2.24	9.00
1200	2771	257.7	0.153	0.618	2.46	9.88
1400	3233	278.4	0.166	0.667	2.66	10.7
1600	3695	297.6	0.177	0.714	2.84	11.4

(Table 5-1 continued on next page)

(Table 5-1 continued)

Orifice (in.)							
$3/15$	$1/4$	$3/8$	$1/2$	$5/8$	$3/4$	$7/8$	1
Orifice (gpm)							
0.641	1.14	2.56	4.56	7.12	10.3	14.0	18.22
0.909	1.61	3.62	6.45	10.1	14.5	19.7	25.8
1.111	1.97	4.44	7.90	12.3	17.8	24.2	31.6
1.28	2.28	5.12	9.12	14.2	20.5	27.9	36.5
1.43	2.54	5.73	10.2	15.9	22.9	31.2	40.7
1.57	2.79	6.28	11.2	17.4	25.1	34.2	44.6
1.70	3.01	6.78	12.1	18.8	27.1	36.9	48.2
1.81	3.22	7.24	12.9	20.1	28.9	39.4	51.5
1.92	3.42	7.69	13.7	21.3	30.8	41.9	54.7
2.03	3.60	8.11	14.4	22.5	32.4	44.1	57.6
2.22	3.94	8.89	15.8	24.7	35.5	48.4	63.1
2.40	4.26	9.59	17.1	26.6	38.4	52.2	68.2
2.55	4.56	10.3	18.3	28.5	41.0	55.9	73.0
2.72	4.83	10.9	19.4	30.2	43.5	59.1	77.4
2.87	5.10	11.5	20.4	31.8	45.9	62.4	81.6
3.21	5.70	12.8	22.8	35.6	51.3	69.8	91.1
3.51	6.24	14.0	24.9	39.0	56.2	76.5	99.8
3.80	6.74	15.2	27.0	42.4	60.7	82.3	108
4.05	7.20	16.2	28.8	45.0	64.9	88.3	115
4.30	7.65	17.2	30.6	47.8	68.7	93.5	122
4.54	8.06	18.1	32.2	50.4	72.5	98.7	129
4.95	8.83	19.8	35.3	55.1	79.4	108	141
5.36	9.54	21.4	38.2	59.6	85.8	117	153
5.73	10.2	22.9	40.7	63.6	91.7	125	163
6.08	10.8	24.3	43.3	67.5	97.3	132	173
6.41	11.4	25.6	45.6	71.2	103	140	183
7.16	12.7	28.6	50.9	79.6	115	156	204
7.86	14.0	31.4	55.9	87.3	126	171	223
8.50	15.1	33.9	60.2	94.3	136	185	241
9.09	16.1	36.2	64.5	101	145	197	258
9.62	17.1	38.4	68.4	107	154	210	274
10.1	18.0	40.5	72.1	112	162	221	288
11.1	19.7	44.4	79.0	123	178	242	316
12.0	21.3	48.0	85.3	133	193	261	341
12.8	22.8	51.2	91.2	142	205	279	365
13.6	24.2	54.4	96.8	151	216	296	387
14.3	25.4	57.2	102	159	229	312	407
15.7	27.9	62.8	112	174	251	342	446
17.0	30.1	67.8	121	188	271	369	482
18.1	32.2	72.4	129	201	289	394	515
19.2	34.2	76.9	137	213	308	419	547
20.3	36.0	81.1	144	225	324	441	576
22.2	39.4	88.9	158	247	355	484	631
24.0	42.6	95.9	171	266	384	522	682
25.5	45.6	103	183	285	410	559	730

(Table 5-1 continued)

psi	Water (ft)	Velocity Through Orifice (fps)	$1/64$	$1/32$	$1/16$	$1/8$
					Flow Through	
1800	4157	315.7	0.188	0.756	3.01	12.1
2000	4619	332.7	0.198	0.798	3.18	12.7
2100	4850	341.0	0.203	0.817	3.25	13.1
2200	5081	349.0	0.208	0.837	3.34	14
2300	5312	356.8	0.212	0.856	3.42	13.7
2400	5543	364.5	0.217	0.875	3.49	14.0
2500	5774	372.0	0.221	0.892	3.55	14.2
3000	6929	407.5	0.242	0.978	3.90	15.6
3500	8083	440.2	0.262	1.05	4.20	16.9
4000	9238	470.6	0.280	1.13	4.50	18.0
4500	10393	499.1	0.297	1.20	4.76	19.1
5000	11548	526.1	0.212	1.26	5.03	20.2
6000	13857	576.3	0.342	1.38	5.50	22.0
7000	16167	622.5	0.370	1.49	5.94	23.8
8000	18476	665.5	0.395	1.59	6.35	25.5
9000	20786	705.8	0.419	1.69	6.74	27.0
10000	23095	744.0	0.442	1.78	7.12	28.5
12000	27714	815.0	0.485	1.95	7.79	28.5
14000	32333	880.3	0.524	2.11	8.42	33.7
16000	36952	941.1	0.560	2.26	9.00	36.0
18000	41571	998.2	0.594	2.40	9.55	38.2
20000	46190	1052	0.626	2.52	10.1	40.4

continued

Orifice (in.)							
$3/15$	$1/4$	$3/8$	$1/2$	$5/8$	$3/4$	$7/8$	1
Orifice (gpm)							
27.2	48.3	109	194	302	435	591	774
28.7	51.0	115	204	318	459	624	816
29.4	52.2	117	209	326	470	639	836
30.1	53.4	120	214	334	481	655	855
30.8	54.6	123	219	341	491	670	875
31.4	55.8	126	223	348	502	684	894
32.1	57.0	128	228	356	513	698	911
35.1	62.4	140	249	390	562	765	998
38.0	67.4	152	270	422	607	823	1078
40.5	72.0	162	288	450	649	883	1152
43.0	76.5	172	306	478	687	935	1222
45.4	80.6	181	322	504	725	987	1290
49.5	88.3	198	353	551	794	1078	1410
53.6	95.4	214	382	596	858	1168	1528
57.3	102	229	407	636	917	1247	1629
60.8	108	243	433	675	973	1321	1730
64.1	114	256	456	712	1025	1395	1822
70.3	125	281	499	780	1123	1529	1998
75.8	135	303	539	843	1212	1650	2158
81.1	144	324	576	901	1298	1765	2305
86.1	153	344	611	956	1376	1870	2445
90.9	161	362	645	1008	1450	1973	2578

tom, causing a float or bob to ascend. The tube is narrow at the bottom, where fluid velocity is greatest, and widest at the top, where velocity is at a minimum. The float indicates the flow rate. The float's specific gravity is slightly greater than that of the fluid. During flow, the float continues to rise until it reaches a point where drag forces are balanced by the weight and buoyancy forces.

The mean fluid velocity can be computed from

$$w_m = \left(\frac{2g\phi}{C_D F}\left(\frac{\rho_B - \rho}{\rho}\right)\right)^{1/2} \tag{5-14}$$

where

$$F = \frac{\pi}{4}((D + \alpha'z)^2 - d^2) \tag{5-15}$$

F = annular flow area
D = diameter of tube at inlet
z = measured vertical distance from inlet
α' = constant related to tube taper (set by manufacturer)
C_D = discharge coefficient

Figure 5-10 gives discharge coefficients for sharp-edged orifices and rotameters.

Sample Calculation 5-1. A pitot tube is used to obtain a point velocity measurement of water flowing through a pipe at 130°F. The manometer fluid is mercury ($\gamma = 13.6$). The manometer reading is 15 inches and the pitot tube coefficient is 0.95. Determine the point velocity of the water.

Solution. Equation (5-132) will be used:

$$w_1 = C\sqrt{2g(\rho_m - \rho)\Delta h/\rho}$$

$$\rho = 61.2 \text{ lb}_m/\text{ft}^3$$

$$\rho_m = (13.60)(62.4 \text{ lb}_m/\text{ft}^3) = 848 \text{ lb}_m/\text{ft}^3$$

$$\Delta h = (15 \text{ in.})(1 \text{ ft}/12 \text{ in.}) = 1.25 \text{ ft}$$

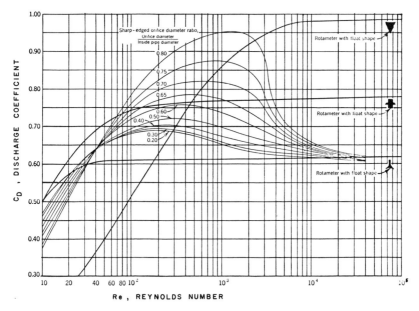

Figure 5-10. Discharge coefficients for rotameters and sharp-edge orifices.[6]

The velocity in English units is

$$w_1 = 0.95\sqrt{\frac{(2)(32.2)(848 - 61.2)(1.25)}{61.2}}$$

$$= 30.6 \text{ ft/sec}$$

Sample Calculation 5-2. A liquid of density 77.2 lbs/ft³ and viscosity 0.74 cp is flowing through a 7.5-in. ID pipe. A sharp-edged orifice with a diameter of 1.14 in. is installed in the pipeline. The measured pressure drop across the orifice is 38 psi 34.6 lb_f/ft². Calculate the volumetric flow rate and the average velocity of the liquid through the pipe.

Solution

$$Q = \frac{C_D \pi}{4} d_0^2 \sqrt{\frac{2(P_1 - P_2)/\rho}{1 - (d_0/d_1)^4}}$$

d_1 = 7.5-in. = 0.63 ft. d_0 = 1.14 in. = 0.1 ft

$$\frac{d_0}{d_1} = \frac{1.14}{7.5} = 0.153$$

Examining the discharge coefficient-Reynolds number plot (Figure 5-10), note that for Re > 20,000, C_D is roughly the same regardless of the diameter ratio. Hence, $C_D \simeq 0.61$ and

$$Q = \frac{0.61(\pi)}{4}(0.10)^2\sqrt{\frac{2 \times 34.6/77.2}{1 - (0.153)^4}g_c}$$

$$= 0.026\frac{\text{ft}^3}{\text{s}}(\text{or } 11.6 \text{ gpm})$$

The average velocity of the liquid is

$$w = \frac{0.026 \text{ ft}^3/\text{s}}{\frac{\pi}{4}(0.63 \text{ ft})^2} = 8.34 \times 10^{-2} \text{ fps}$$

Re is calculated to check if it is greater than 20,000 for $C_D = 0.61$.

$$\mu = 0.74 \times 10^{-3}\text{kg/m-sec} = (1.54 \times 10^{-5}\text{lb}_f\text{-s/ft}^2)$$

$$\text{Re} = \frac{d_1 w \rho}{\mu} = \frac{(0.63)(8.34 \times 10^{-2})(77.2)}{1.54 \times 10^{-5}}$$

$$= 2.634 \times 10^5$$

Since the Reynolds number is greater than 20,000, a good value for the discharge coefficient was selected.

Sample Calculation 5-3. A Venturi meter is used to measure air flow through a pipe. The upstream diameter of the Venturi is 4.5 inches (d_1), and the throat diameter ($d_0 = d_2$) is 1.75 in. The pressure drop across the meter is 0.290 psi, and the gas density is 0.07 lb_m/ft^3.

Solution. From continuity,

$$Q = F_1 w_1 = F_2 w_2$$

where Q is the volumetric discharge. Thus,

$$\frac{\pi}{4}\left(\frac{4.5}{12}\right)^2 w_1 = \frac{\pi}{4}\left(\frac{1.75}{12}\right)^2 w_2$$

or

$$Q = 0.1104 w_1 = 0.0167 w_2$$

Applying the Bernoulli equation for $z_1 = z_2$,

$$P_1 - P_2 = 0.290 \times 144 = 41.76 \text{ lb}_f/\text{ft}^2$$

$$\frac{P_1 - P_2}{\rho} = \frac{w_2^2}{2g} - \frac{w_1^2}{2g}$$

$$\frac{41.76}{0.07} = \frac{Q^2}{2g}\left[\left(\frac{1}{.0167}\right)^2 - \left(\frac{1}{0.1104}\right)^2\right]$$

Solving for the volumetric discharge gives $Q = 10.97$ cfs (or 2764 lb_m/hr) of air.

References

1. ASME Research Committee, *Fluid Meters: Their Theory and Application,* The American Society of Mechanical Engineers, New York (1959).
2. Cheremisinoff, N. P., *Applied Fluid Flow Measurement,* Marcel Dekker, Inc., New York (1979).
3. Cheremisinoff, N. P., *Process Level Instrumentation and Control,* Marcel Dekker, Inc., New York (1981).
4. Cheremisinoff, N. P. and Azbel, D., *Fluid Mechanics and Unit Operations,* Ann Arbor Science Pub., Ann Arbor, MI (1983).
5. Cheremisinoff, N. P., *Fluid Flow: Pumps, Pipes and Channels,* Ann Arbor Science Pub., Ann Arbor, MI (1981).
6. Brown, G. et al., *Unit Operations,* John Wiley & Sons Inc., New York (1950).

6

FLOW
THROUGH COILS AND
AROUND TUBES

Flow Through Coils

Coils are used in a variety of process equipment. They have critical Reynolds numbers (Re_{cr}) that greatly differ from those encountered in straight pipes. Here, the friction factor is a function of the ratio of coil diameter D to pipe diameter d. Typical critical values for different D/d ratios are given as follows:

D/d	15.5	18.7	50	2050
Re_{cr}	7600	7100	6000	2270

These values show that the smaller the coil diameter, the higher the critical value of the Reynolds number and, consequently, the longer the flow remains in the laminar regime (that is, the critical Re of 2100 marking the transition between laminar and turbulent flows through pipes is greatly exceeded for coils). The resistance of laminar flow in coils may be estimated from the Darcy-Weisbach equation

$$\left(h_\ell = -\frac{\Delta P}{\gamma} = \lambda \frac{L}{D} \frac{w^2}{2g} \right)$$

using the following empirical friction factor:

$$\lambda = C' \frac{64}{Re} \tag{6-1}$$

Coefficient C' approaches unity with increasing coil radius of curvature. C' may be computed from the following formula or values:

$$C' = \cfrac{1}{1 - \left[1 - \left(\cfrac{11.6}{Re\sqrt{d/D}}\right)^{0.45}\right]^{1/0.45}} \qquad (6\text{-}2)$$

$Re\sqrt{d/D}$	10	50	100	250	400	600	1000	2000
C'	1.0	1.2	1.5	2.0	2.5	3.0	4.5	5.0

For turbulent flow through coils of $D/d > 500$, resistances are comparable to those in straight pipes. At very high Reynolds numbers ($Re > 110,000$), the resistance coefficient λ is independent of Re and can be expressed by

$$\lambda = 0.0238 + 0.0891 \ d/D \qquad (6\text{-}3)$$

To estimate hydraulic resistances through coils, assume the coil to be a return bend and apply the following equation to obtain an L_e/D ratio:

$$\frac{L_e}{D} = 0.0202 X' a^{1.10} Re^{0.032} \qquad (6\text{-}4)$$

where a is the bending angle in grades and X' an empirical function of the ratio of bending radius R to pipe radius r. Values of X' are given as follows:

R/r	2	4	6	8	10	20	32
X'	4	1.2	1.2	1.7	2.2	4.8	7.6

Flow Normal to Tube Banks

Flow normal to a bank of tubes is encountered in a variety of industrial equipment such as tubular heat exchangers and condensers. The magnitude of the flow resistance in such equipment is due to the contraction and expansion of the flow. This resistance also depends on the tube layout (e.g., staggered or unstaggered) relative to the flow direc-

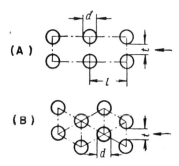

Figure 6-1. Flow normal to a bank of tubes. Flow resistance is a function of the tube layout: (a) an unstaggered layout, (b) a staggered layout.

tion as illustrated in Figure 6-1. The critical Reynolds number for these flows can be estimated from

$$Re_{cr} = \frac{t' w_{max} \gamma}{\mu g} \simeq 40 \tag{6-5}$$

where t' is the distance between tubes and w_{max} is the maximum fluid velocity (in the narrow section).

For turbulent flow (i.e., $Re > 40$), flow resistance may be estimated from

$$h_r = \lambda N' \frac{w_{max}^2}{2g} \tag{6-6}$$

where: $h_r = \Delta P/\gamma$ = resistance of the liquid column
N' = number of tube rows in the direction of flow
λ = resistance coefficient depending on Re and the tube layout

The coefficient λ may be determined from the following formulas. For a single row of tube layout,

$$\lambda = \left[0.175 + \frac{0.32\frac{\ell}{d}}{(t'/d)^n} \right] Re_d \tag{6-7}$$

where e, d, t' are defined in Figure 6-1, and $Re_d = t'w_{max}\gamma/\mu g/ = dw_{max}\gamma/\mu g$.

At $Re_d \leq 10{,}000$ (but still turbulent), λ is constant. Note that when e = 2d, t' = d and Equation 6-7 simplifies to

$$\lambda = 1.82\ Re^{-0.15} \tag{6-8}$$

where $Re = Re_d$. For rows of staggered tubes in the Reynolds number range of 2000–4000, the following equation can be used:

$$\lambda = \left[0.92 + \frac{0.44}{(t/d)^{1.08}}\right]Re_d^{-0.15} \tag{6-9}$$

Or, for the case of t = d,

$$\lambda = 1.40\ Re^{-0.15} \tag{6-10}$$

Equations 6-7 through 6-10 give good estimates when the number of rows, N', exceeds 10. At N' < 10, actual head losses are somewhat higher than those predicted by applying these equations. At N' = 4, the error in applying these correlations is 7%; at N' = 3, 15%; and at N' = 2, up to 30%.

For flow around a single row of tubes (N' = 1), use the following table to obtain values of λ for different t/d ratios at Re = 10,000 (note that λ is constant at Re $\leq$ 10,000 for a specific t/d).

t'/d	0.2	0.5	1.0	1.5	2.0	5.0	10.0
λ	0.64	0.50	0.40	0.36	0.30	0.18	0.10

For laminar flow around tubes (Re < 40), the resistance may be calculated from an expression analogous to the Darcy equation

$$h_r = \lambda\frac{\ell}{d_{eg}}N'\frac{w_{max}^2}{2g} \tag{6-11}$$

where: $h_r = \Delta P/\gamma$ (the resistance of the liquid column)
ℓ = distance between rows in the direction of flow
w_{max} = maximum velocity in the intertubular space
d_{eq} = equivalent diameter equal to four times hydraulic radius

For flow between two rows of tubes, a unit length of tube is $\ell(t + d) - \pi d^2/4$, where $\pi d^2/4$ is part of the volume occupied by neighboring tube segments. The surface is πd and the equivalent diameter is

$$d_{eq} = \frac{4\ell(t' + d) - \dfrac{\pi d^2}{4}}{\pi d} \frac{4\ell(t' + d)}{\pi d} - d \tag{6-12}$$

The equivalent Reynolds number is

$$Re_{eq} = \frac{d_{eq}w_{max}\gamma}{\mu g} \tag{6-13}$$

For a row of staggered tubes, $Re_{eq} = 1 \sim 100$

$$\lambda = \frac{106}{Re_{eq}} \tag{6-14}$$

Substituting this expression for λ into Equation 6-11 gives

$$h_r = \frac{53\mu N' \ell w_{max}}{\gamma d_{eq}^2} \tag{6-15}$$

For a row of unstaggered tubes, λ values for turbulent flow should be multiplied by 1.5.

7

CHANNEL
FLOWS

Flow Through Noncircular Sections

Equivalent diameter is defined as four times the hydraulic radius r_h, where r_h is the ratio of the cross-sectional area of flow and the wetted perimeter (i.e., the perimeter of the channel contacting the fluid):

$$d_{eq} = 4r_h \qquad (7\text{-}1)$$

and

$$r_h = \frac{s}{p'} \qquad (7\text{-}2)$$

where: s = cross-sectional area of channel
p' = wetted perimeter of the channel

Table 7-1 gives specific formulas for the hydraulic radius for channels of various configurations.

For an elliptical cross section, coefficient k in Table 7-1 must be computed. K is a function of the ratio $\acute{S} = (a - b)/(a + b)$. Values are given as follows:

$\acute{S}$:	0.2	0.3	0.4	0.5	0.6	0.7	0.8	0.9	1.0
K:	1.010	1.023	1.040	1.064	1.092	1.127	1.168	1.216	1.273

There are two approaches to evaluating resistances for laminar flows through irregular configurations (full channel flows). In the first

Table 7-1
Hydraulic Radii for Different Channel Configurations[1]

Cross Section	r_h
Circular pipe, diameter D	$D/4$
Annulus between two concentric pipes; D and d are the outside and inside diameters of the annulus	$\dfrac{D - d}{4}$
Rectangular duct with sides a and b	$\dfrac{ab}{2(a + b)}$
Square duct with a side a	$a/4$
Ellipse with axes a and b	$\dfrac{ab}{K(a + b)}$
Semicircle of diameter D	$D/4$
A shallow flat layer with depth h	h
Liquid film with thickness t on the vertical pipe of diameter D	$t - \dfrac{t^2}{D} \simeq t$

method, the Reynolds number is evaluated from an equivalent diameter defined by Equations 7-1 and 7-2. A friction factor can then be computed from an expression of the following form:

$$\lambda = \frac{a}{Re} \tag{7-3}$$

The value of a depends on the flow geometry. Othmer[2] gives values for different configurations which are noted in Table 7-2.

The second method for estimating losses in the laminar regime for noncircular cross sections is based on the derivation of Poiseuille-type equations. As an example, consider the flow configuration of an annular cross section. Following the derivation of Poiseuille's law (see Cheremisinoff et al.[1] for details), an expression for the resistance per unit length of channel is

$$- \frac{dP}{dL} = \frac{32\mu\overline{w}}{D^2 + d^2 - \dfrac{D^2 - d^2}{\ln D/d}} \qquad (7\text{-}4)$$

where $\overline{w}$ is the average superficial velocity. For a rectangular section of sides a and b, Poiseuille's equation is

$$- \frac{dP}{dL} = \frac{4\overline{w}\mu}{abn} \qquad (7\text{-}5)$$

n is a function of a/b; typical values are given as follows:

a/b:	0.1	0.2	0.3	0.4	0.5	0.6	0.7	0.8	0.9	1.0
n:	0.03	0.06	0.08	0.10	0.11	0.133	0.136	0.138	0.139	0.140

For an elliptical cross section, the head loss is

$$- \frac{dP}{dL} = \frac{4\overline{w}\mu(a^2 + b^2)}{a^2 b^2} \qquad (7\text{-}6)$$

Table 7-2
Equivalent Diameters and Values for
Parameter a in Equation 7-3

Cross section	h/b	D_{eq}	a
Circular, with diameter D	–	D	64
Square with side h	–	h	57
Triangle with side h	–	0.58h	53
Ring with width b	–	2b	96
	0.7	1.17h	65
	0.5	1.30h	68
Ellipse with axes h and b	0.3	1.44h	73
(h = small axis)	0.2	1.50h	76
	0.1	1.55h	78
	1/∞	2h	96
	0.1	1.82h	85
	0.2	1.67h	76
Rectangle with sides h and b	0.25	1.60h	73
(h = small side)	0.33	1.50h	69
	0.50	1.33h	62

where a and b are the semiaxes of an ellipse.

A limiting case is that of two infinitely wide plates separated by a distance of 2b. Poiseuille's equation may be written as

$$-\frac{dP}{dL} = \frac{3\mu\overline{w}}{b^2} \qquad (7\text{-}7)$$

The velocity distribution between the two parallel plates is given by the following equation:

$$w = \frac{3}{4}\frac{\Gamma(b^2 - y^2)}{b^3\gamma} \qquad (7\text{-}8)$$

where: Γ = weight rate per unit plate width
   y = distance to the plane of symmetry (i.e., the plane in the middle between the plates)
   γ = specific gravity

Flow Formulas

One of the best-known formulas for open channel flow is the Chezy formula:

$$w^2 = \frac{W}{K}\frac{A}{p'}S \qquad (7\text{-}9)$$

This formula is most often written as

$$w = C\sqrt{h_D S} \qquad (7\text{-}10)$$

where: h_D = hydraulic mean depth = A/p'
   C = Chezy constant
   S = slope of the fluid's free surface

Chezy's constant can be evaluated from the following formula:

$$C = \frac{23 + \dfrac{0.00155}{S} + \dfrac{1}{\epsilon}}{1 + \left(23 + \dfrac{0.00155}{S}\right)\dfrac{\epsilon}{\sqrt{h_D}}} \qquad (7\text{-}11)$$

Table 7-3
Typical Values for Roughness Coefficients
for Open-Channel Flows[3]

Channel Description	Range of ϵ (in.)
Steel Channels	0.011–0.017
Planned Timber, Joints Flush	0.010–0.014
Sawn Timber, Joints Uneven	0.011–0.015
Concrete	0.011–0.017
Brickwork	0.012–0.018
Excavated Channels (earth)	0.016–0.030
(gravel)	0.022–0.030
(rock cut, smooth)	0.023–0.040
(rock cut, jagged)	0.035–0.060
Natural Channels (clean, regular)	0.025–0.040
(rocky, brushwood, debris)	0.050–0.150
Floodplains (short grass pasture)	0.025–0.035
(brushwood)	0.035–0.070

where ϵ is a coefficient of roughness for the channel material.

A second empirical formula used for open-channel flows is the Manning equation:

$$w = \frac{1}{\epsilon} h_D^{2/3} S^{1/2} \text{ (in SI units)} \tag{7-12}$$

or

$$w = \frac{1.49}{\epsilon} h_D^{2/3} S^{1/2} \text{ (in English units)} \tag{7-13}$$

For accuracy, the Manning formula is preferred.

Unlike full-pipe flow, the roughness element ϵ for open-channel flows is sensitive to the hydraulic state of the channel. Table 7.3 gives typical values for ϵ for different straight-channel conditions. For non-straight channels, ϵ values should be increased by 30%.[3]

There are three general flow cases that are encountered frequently: (1) uniform flows, (2) nonuniform flows, and (3) unsteady flows in open channels. Each of these systems, along with the general expressions for estimating volumetric flows, are outlined here.

For uniform, fully developed flow in open channels, the Chezy and Manning formulas provide adequate estimates of the mean fluid velocity from which a volumetric discharge rate can be computed. From the Manning equation, the volumetric discharge rate is

$$Q = \frac{1}{\epsilon} A h_D^{2/3} S^{1/2} \qquad (7\text{-}14)$$

where A is the flow area.

The group $A h_D^{2/3}$ is called the *section factor* and contains information on the geometry of the cross section. The group $1/\epsilon A h_D^{2/3}$ is referred to as the *conveyance* and contains information on both the channel geometry and the channel's relative roughness.

For a rectangular flow cross section, the depth corresponding to uniform flow for a specified discharge and bed slope is constant at all points and is referred to as the normal depth. For nonrectangular channels of uniform flow, normal depth is defined as the maximum depth in the cross section. The normal depth meets the criterion of

$$\frac{dh}{dx} = 0 \qquad (7\text{-}15)$$

For a wide channel ($b \gg h_o$, where b is the channel width), the normal depth can be estimated directly from the Chezy formula:

$$Q = AC\sqrt{h_o S} = bh_o C\sqrt{h_o S} \qquad (7\text{-}16)$$

The hydraulic mean depth approaches the depth for a wide channel

$$h_o = \sqrt[3]{\frac{Q^2}{b^2 S C^2}} \qquad (7\text{-}17)$$

or

$$h_o = \sqrt[3]{q^2/SC^2} \qquad (7\text{-}18)$$

where q is the discharge per unit width (Q/b).

The Manning formula gives the following relation for wide channels:

$$h_o = \left(\frac{\epsilon q}{S^{1/2}}\right)^{3/5} \qquad (7\text{-}19)$$

For the more general case in which h_o is comparable to channel width b, the hydraulic mean depth (a definition analogous to hydraulic radius in pipe flow) may be used:

$$h_o = \frac{bh_o}{b + 2h_o} \tag{7-20}$$

Combining this with the Manning relationship,

$$Q = \frac{1}{\epsilon}(bh_o)(h_D)^{3/2}\sqrt{S} \tag{7-21}$$

Note that if uniform flow does not exist in the channel, there may be several possible fluid levels for a specified discharge. A trial-and-error solution is then required for Equation 7-21.

For nonrectangular channels, the following procedure is useful for determining the normal depth in a channel for uniform flow, with the Manning formula rewritten as:

$$\frac{Q}{\sqrt{S}} = \frac{1}{\epsilon}Ah_D{}^{2/3} = K' \tag{7-22}$$

Step 1. Compute values of K' for a suitable range, where K' is the conveyance of h (up to the maximum depth in the cross section).
Step 2. Prepare a plot of conveyance K' versus depth.
Step 3. Using the plot, read the normal depth for the K' value corresponding to the specified Q and S values.

Channel geometry affects the mean hydraulic depth, which, in turn, affects not only the volumetric flow, but the economics of the design as well. The most economical channel design is based on a geometry that requires the minimum flow cross section for a specified discharge rate. From the Manning formula, this becomes an exercise in maximizing the term A^5/p'^2, or since A is often fixed, p' is optimized. Obviously, a semicircular channel will provide the *best hydraulic section;* however, this often poses fabrication problems.

The two most common geometries employed are rectangular and trapezoidal channels. Table 7.4 gives formulas for the cross-sectional

Table 7-4
Formulas for Rectangular and Trapezoidal Channels

Channel Description	Cross-Sectional Area of Flow	Wetted Perimeter	Best Hydraulic Depth
Rectangular Channel	$A = bh$	$p' = 2h + b$ $= 2h + \dfrac{A}{h}$	$H_D = \dfrac{bh}{2h + b} = \dfrac{2h^2}{4h}$ $= h/2$
Trapezoidal Channel	$A = h(zh + b')$ $= h(zh + p' - 2h\sqrt{(z^2 + 1)})$	$p' = 2h\sqrt{(z^2 + 1)} + b'$ where z is the side slope; if z is a variable and 'A' and 'h' are fixed, then — $p' = \dfrac{A}{h} - zh + 2h\sqrt{(z^2 + 1)}$	$H_D = h/2$; with $\theta = 60°$ for the best hydraulic section

area of flow, wetted perimeter, and the best hydraulic depth for these two geometries. The optimum hydraulic depth is determined by minimizing p' with respect to depth (i.e., dp'/dh = 0).

To estimate free surface profiles for gradually varied flows, the methods of Sellin[4] and Henderson[5] are best. The reader is referred to these sources for details. The following assumptions are used in their analysis:

1. If the slope of the channel bed S is small, the depth h may be considered a normal distance.
2. The velocity distribution across the channel is uniform, i.e., the results for uniform flow are applied locally to gradually varied flow.
3. The total energy head H and the bed elevation Z are considered positive above a horizontal datum.
4. The relationship between change in depth and change in cross-sectional area can be approximated by dA/dh = b, where b is the channel width.

The general equation of gradually varied flow is

$$\frac{dh}{dx} = \frac{S - j}{1 - \dfrac{Q^2 b}{gA^3}} \tag{7-23}$$

where j is the slope of the total energy line.

For a rectangular cross-sectional channel,

$$\frac{dh}{dx} = \frac{S - j}{1 - q^2/gh^3} \tag{7-24}$$

where: $A = bd$

$q = Q/b$

From assumption 2, the Manning formula is used to obtain an expression for j:

$$j = \frac{w^2 \epsilon^2}{h_{Dm}^{4/3}} \tag{7-25}$$

where h_{Dm} is the hydraulic mean depth. Substituting into Equation 7-24, an expression for a rectangular channel is obtained:

$$\frac{dh}{dx} = \frac{S - \dfrac{w^2}{C^2 h_{Dm}}}{1 - w^2/gh} \tag{7-26}$$

The gradually varied flow expression may be solved to obtain the free-surface profile, that is, the fluid depth as a function of x along the channel. The general expression, Equation 7-23, can be rearranged to solve for x:

$$x = \int dx = \int \left(\frac{1 - B}{S - d}\right) dh \tag{7-27}$$

where:

$$B = \frac{Q^2 b}{gA^3} \text{ and } d = \frac{\epsilon^2 Q^2}{A^2 h_{Dm}^{4/3}}$$

Note that B and d are functions of h (since A, b, h_{Dm} are functions of h). Hence, Equation 7-27 is $x = \int f(h) dh$. If f(h) is a simple function, the integral may be evaluated directly.

Simpson's rule may be used to evaluate quadratic functions numerically. In applying Simpson's rule, Equation 7-27 is treated in the form $dx/dh = f(h)$, whereby values of dx/dh corresponding to a chosen range of h values are first computed. Simpson's rule is then applied to evaluating the integral

$$\Delta x = \int_{h1}^{h} \left(\frac{dx}{dh}\right) dh \tag{7-28}$$

whereby Δx is a distance along the channel between specified depths h_1 and h_2, i.e., Equation 7-28 gives the area under the curve between the coordinates h_1 and h_2 for the curve dx/dh versus h. A quadratic is fitted to match values at h_o, h_1, h_2, h_3, h_4, and so forth. The number of intervals must be even, and the area under each successive three points is approximated by the area under the quadratic

$$\text{Area under curve} = \frac{h_n - h_o}{3n}[f(h_o) + 4f(h_1) + 2f(h_2) + 4f(h_3) \\ + 2f(h_4) + \ldots + 4f(h_{n-1}) + f(h_n)] \tag{7-29}$$

Weirs and Flumes

The three most common weir designs are shown in Figure 7-1. The basis of measurement is the fluid level at a given distance upstream from the weir, which is proportional to the flow. The form of the crest is important for accurate measurements. A problem associated with rectangular weirs for water flow is that the flow will be contracted as it passes over the weir. Hence, the effective width of the weir is smaller than the crest. Cipolletti weirs are designed with sloping sides to compensate for this contraction.

The design formula for flow over a weir (in English units) is

$$q = C_W^{2/3} \sqrt{2g}\, H^{3/2} \qquad\qquad (7\text{-}30)$$

where: H = upstream head above the crest
   q = flow per unit width (cfs/ft)
   C_W = nonuniformity coefficient to compensate for the non-uniformity of flow ($C_W \le 1$)

Permanently installed weirs should be calibrated after installation to determine the proper coefficient.

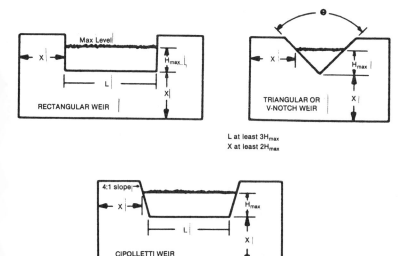

Figure 7-1. Common types of sharp-crested weirs.

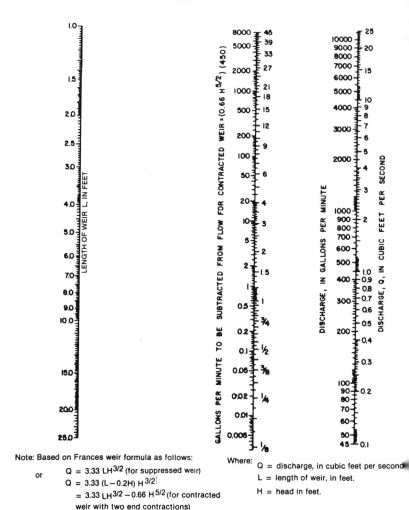

Note: Based on Frances weir formula as follows:

or
$Q = 3.33\ LH^{3/2}$ (for suppressed weir)

$Q = 3.33\ (L - 0.2H)\ H^{3/2}$

$= 3.33\ LH^{3/2} - 0.66\ H^{5/2}$ (for contracted weir with two end contractions)

Where:
Q = discharge, in cubic feet per second
L = length of weir, in feet.
H = head in feet.

Figure 7-2. Nomograph for estimating the discharge over rectangular weirs.

Rectangular weirs are of the straight or notched variety. The former is called a suppressed weir without end contractions. A notched weir may have one or two end contractions. All engineering weir formulas have as their basis the Francis formula. Figure 7-2 is a nomograph of the Francis formula, which can be used for a suppressed weir or a weir with standard end contractions. The nomograph should not be applied for discharges with very low heads, which can cause the nappe of the fluid to cling to the weir face.

For measuring low flows (less than 0.03 m³/s), a V-notched weir is preferred. A nomograph for estimating flowrates over 60° and 90° triangular weirs is given in Figure 7-3.

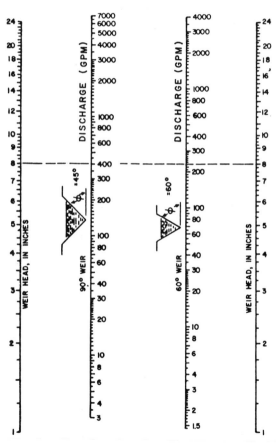

Figure 7-3. Nomograph for estimating flowrates over 60° and 90° triangular weirs.

The formulas used for the nomographs assume the fluid velocity of approach is small. When this is not the case, a correction factor $w^2/2g$ should be added to the head term, where w is the approach velocity. If the crest height is too high, the flow may occur over the full width of the weir and problems may develop when a vacuum forms under the nappe. For rectangular weirs, the crest height should not exceed 5H.

Tables 7-5 and 7-6 give discharge capacities for water flow over rectangular weirs with end contractions and triangular notch weirs, respectively.

Parshall flumes are the most common devices for measuring flows in sewers. The basic configuration consists of three parts: (1) a converging section, (2) a throat section, and (3) a diverging section. Dimensions and capacities of Parshall flumes are given in Figure 7-4. Flow measurement is based on the fluid surface in the converging section. The flow level normally is measured back from the crest of the flume at a distance of two-thirds the length of the converging section. Measurements should always be made in a stilling well instead of the flume itself, because sudden flow changes are dampened in stilling wells. Figure 7-5 estimates flow capacities for Parshall flumes.

The Palmer-Bowlus flume is the simplest design, which, for sewer flows, consists of a level section of floor placed into a sewer. The length of the floor is sized at the same diameter as the conduit. Figure 7-6 shows various configurations of Palmer-Bowlus flumes. Flow through a Palmer-Bowlus flume can be computed from the following formulas (in English units):

$$\frac{Q^2}{g} = \frac{Ac^3}{b} \tag{7-31}$$

and

$$\frac{w_c^2}{2g} = \frac{A_c}{2b} = \frac{h_c}{2} \tag{7-32}$$

where: A_c = area of critical depth
h_c = critical depth
w_c = critical velocity
b = width of flume

Figure 7-7 provides approximate head losses for water flows in weirs and flumes.

(Text continued on page 121)

Table 7-5
Discharge Rates from Rectangular Weirs with End Contractions

Head (H) in Inches	Length (L) of weir in feet				Head (H) in Inches	Length (L) of weir in feet		
	1	3	5	Additional g.p.m. for each ft. over 5 ft.		3	5	Additional g.p.m. for each ft. over 5 ft.
1	35.4	107.5	179.8	36 05	8	2338	3956	814
1¼	49.5	150.4	250.4	50.4	8¼	2442	4140	850
1½	64.9	197	329 5	66.2	8½	2540	4312	890
1¾	81	248	415	83.5	8¾	2656	4511	929
2	98.5	302	506	102	9	2765	4699	970
2¼	117	361	605	122	9¼	2876	4899	1011
2½	136.2	422	706	143	9½	2985	5098	1051
2¾	157	485	815	165	9¾	3101	5288	1091
3	177.8	552	926	187	10	3216	5490	1136
3¼	199.8	624	1047	211	10½	3480	5940	1230
3½	222	695	1167	236	11	3716	6355	1320
3¾	245	769	1292	261	11½	3960	6780	1410
4	269	846	1424	288	12	4185	7165	1495
4¼	293 6	925	1559	316	12½	4430	7595	1575
4½	318	1006	1696	345	13	4660	8010	1660
4¾	344	1091	1835	374	13½	4950	8510	1780
5	370	1175	1985	405	14	5215	8980	1885
5¼	395.5	1262	2130	434	14½	5475	9440	1985
5½	421.6	1352	2282	465	15	5740	9920	2090
5¾	449	1442	2440	495	15½	6015	10400	2165
6	476.5	1535	2600	528	16	6290	10900	2300
6¼		1632	2760	560	16½	6565	11380	2410
6½		1742	2920	596	17	6925	11970	2520
6¾		1826	3094	630	17½	7140	12410	2640
7		1928	3260	668	18	7410	12900	2745
7¼		2029	3436	701.5	18½	7695	13410	2855
7½		2130	3609	736	19	7980	13940	2970
7¾		2238	3785	774	19½	8280	14460	3090

This table is based on Francis formula:

$$Q = 3.33 \ (L-0.2H) \ H^{1.5}$$

where

Q = cu. ft. of water flowing per second.
L = length of weir opening in feet. (should be 4 to 8 times H).
H = head on weir in feet (to be measured at least 6 ft. back of weir opening).
a = should be at least 3 H.

Table 7-6
Discharge Rates from Triangular Notch Weirs
with End Contractions

Head (H) in Inches	Flow in Gallons Per Min. 90° Notch	60° Notch	Head (H) in Inches	Flow in Gallons Per Min. 90° Notch	60° Notch	Head (H) in Inches	Flow in Gallons Per Min. 90° Notch	60° Notch
1	2.19	1.27	6¾	260	150	15	1912	1104
1¼	3.83	2.21	7	284	164	15½	2073	1197
1½	6.05	3.49	7¼	310	179	16	2246	1297
1¾	8.89	5.13	7½	338	195	16½	2426	1401
2	12.4	7.16	7¾	367	212	17	2614	1509
2¼	16.7	9.62	8	397	229	17½	2810	1623
2½	21.7	12.5	8¼	429	248	18	3016	1741
2¾	27.5	15.9	8½	462	267	18½	3229	1864
3	34.2	19.7	8¾	498	287	19	3452	1993
3¼	41.8	24.1	9	533	308	19½	3684	2127
3½	50.3	29.0	9¼	571	330	20	3924	2266
3¾	59.7	34.5	9½	610	352	20½	4174	2410
4	70.2	40.5	9¾	651	376	21	4433	2560
4¼	81.7	47.2	10	694	401	21½	4702	2715
4½	94.2	54.4	10½	784	452	22	4980	2875
4¾	108	62.3	11	880	508	22½	5268	3041
5	123	70.8	11½	984	568	23	4565	3213
5¼	139	80.0	12	1094	632	23½	5873	3391
5½	156	89.9	12½	1212	700	24	6190	3574
5¾	174	100	13	1337	772	24½	6518	3763
6	193	112	13½	1469	848	25	6855	3958
6¼	214	124	14	1609	929			
6½	236	136	14½	1756	1014			

Based on formula:

$$Q = (C) \, (4/15) \, (L) \, (H) \sqrt{2gH}$$

in which Q = flow of water in cu. ft. per sec.

 L = width of notch in ft. at H distance above apex.

 H = head of water above apex of notch in ft.

 C = constant varying with conditions, .57 being used for this table.

 a = should be not less than ¾ L.

For 90° notch the formula becomes

$$Q = 2.4381 H^{5/2}$$

For 60° notch the formula becomes

$$Q = 1.4076 \, H^{5/2}$$

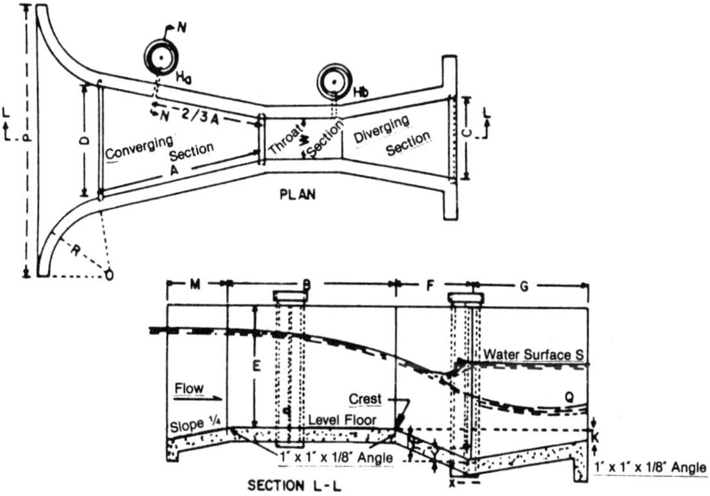

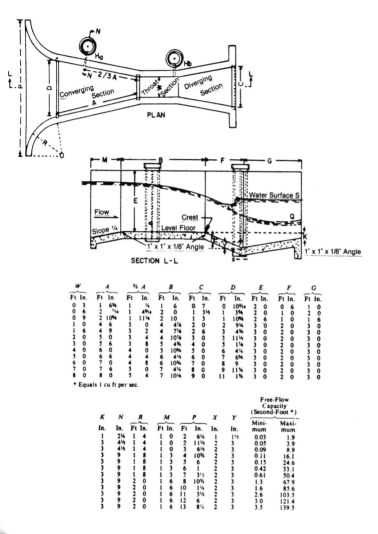

W		A		⅔ A		B		C		D		E		F		G	
Ft	In.	Ft	In.	Ft	In.	Ft	In.	Ft	In.	Ft	In.	Ft	In.	Ft	In.	Ft	In.
0	3	1	6⅜	1	¼	1	6	0	7	0	10³⁄₁₆	2	0	0	6	1	0
0	6	2	⁷⁄₁₆	1	4⁵⁄₁₆	2	0	1	3½	1	3⅝	2	0	1	0	2	0
0	9	2	10⅝	1	11⅛	2	10	1	3	1	10⅝	2	6	1	0	1	6
1	0	4	6	3	0	4	4⅞	2	0	2	9¼	3	0	2	0	3	0
1	6	4	9	3	2	4	7⅞	2	6	3	4⅜	3	0	2	0	3	0
2	0	5	0	3	4	4	10⅞	3	0	3	11½	3	0	2	0	3	0
3	0	5	6	3	8	5	4¾	4	0	5	1⅞	3	0	2	0	3	0
4	0	6	0	4	0	5	10⅝	5	0	6	4¼	3	0	2	0	3	0
5	0	6	6	4	4	6	4½	6	0	7	6⅝	3	0	2	0	3	0
6	0	7	0	4	8	6	10⅜	7	0	8	9	3	0	2	0	3	0
7	0	7	6	5	0	7	4¼	8	0	9	11⅜	3	0	2	0	3	0
8	0	8	0	5	4	7	10⅛	9	0	11	1¾	3	0	2	0	3	0

• Equals 1 cu ft per sec.

K	N	R		M		P		X	Y	Free-Flow Capacity (Second-Foot •)	
In.	In.	Ft	In.	Ft	In.	Ft	In.	In.	In.	Minimum	Maximum
1	2¼	1	4	1	0	2	6¼	1	1½	0.03	1.9
3	4½	1	4	1	0	2	11½	2	3	0.05	3.9
3	4½	1	4	1	0	3	6½	2	3	0.09	8.9
3	9	1	8	1	3	4	10¾	2	3	0.11	16.1
3	9	1	8	1	3	5	6	2	3	0.15	24.6
3	9	1	8	1	3	6	1	2	3	0.42	33.1
3	9	1	8	1	3	7	3½	2	3	0.61	50.4
3	9	2	0	1	6	8	10¾	2	3	1.3	67.9
3	9	2	0	1	6	10	1¼	2	3	1.6	85.6
3	9	2	0	1	6	11	3½	2	3	2.6	103.5
3	9	2	0	1	6	12	6	2	3	3.0	121.4
3	9	2	0	1	6	13	8¼	2	3	3.5	139.5

LEGEND:

- W Size of flume, in inches or feet.
- A Length of side wall of converging section.
- ⅔ A Distance back from end of crest to gage point.
- B Axial length of converging section.
- C Width of downstream end of flume.
- D Width of upstream end of flume.
- E Depth of flume.
- F Length of throat.
- G Length of diverging section.
- K Difference in elevation between lower end of flume and crest.
- N Depth of depression in throat below crest.
- R Radius of curved wing wall.
- M Length of approach floor.
- P Width between ends of curved wing walls.
- X Horizontal distance to H_b gage point from low point in throat.
- Y Vertical distance to H_b gage point from low point in throat.

Figure 7-4. Dimensions and capacities of Parshall flumes.[5]

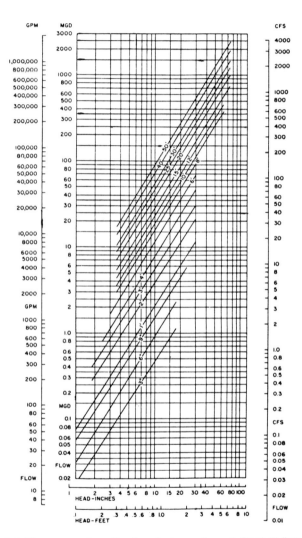

Figure 7-5. Flow curves for estimating discharge through Parshall flumes.[6]

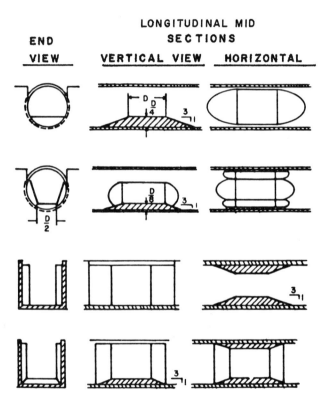

Figure 7-6. Various configurations of Palmer-Bowlus flumes.

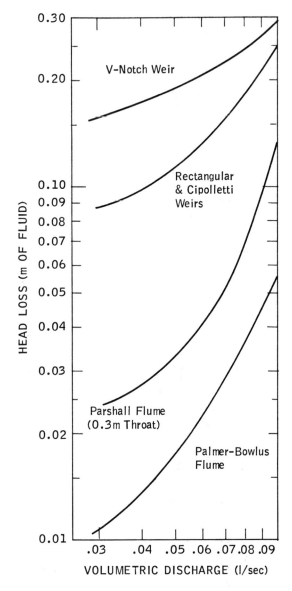

Figure 7-7. Relative head losses for water flows in different types of weirs and flumes.

References

1. Cheremisinoff, N. P. and Azbel, D., *Fluid Mechanics and Unit Operations,* Ann Arbor Science Publishers, Ann Arbor, MI (1983).
2. Othmer, D. F., *Ind. Eng. Chem.,* 37(11): 1112 (1945).
3. Cheremisinoff, N. P., and Gupta, R. (Editors), *"Open-Channel Flows,"* Chapter 13, pp. 329–368, *Handbook of Fluids in Motion,* Ann Arbor Science Pub., Ann Arbor, MI (1983).
4. Sellin, R. H., *Flow in Channels,* McMillan St. Martins Press, London, England (1969).
5. Henderson, F. M., *Open Channel Flows,* Collier Publishers, London, England (1966).
6. *Handbook for Monitoring Industrial Wastewater,* U.S. Environmental Protection Agency, Washington, DC (1973).

8

TWO-PHASE (GAS-LIQUID) FLOWS

Flow Regimes

The term flow regime or flow pattern refers to the nature of the gas-liquid interface in two-phase flows. Typical flow patterns encountered in horizontal pipes are shown in Figure 8-1.

Stratified flow is observed at low gas and liquid flowrates. The liquid flows along the pipe bottom and gas along the top. At high gas flowrates, waves are generated at the interface (called *wavy flow*).

At higher liquid flowrates but low gas flowrates, the pattern is described as *bubble flow.* Small gas bubbles tend to flow in the upper portion of the pipe. At higher gas flowrates, large bullet-shaped bubbles, called plugs, are formed. In *plug flow,* the plugs move through the liquid along the upper portion of the pipe.

At even higher gas flows, *slug flow* exists. In slug flow, frothy slugs of liquid move across the upper portion of the pipe. A wavy layer of liquid exists between the slugs at the bottom of the pipe.

Annular flow exists at still higher gas velocities. Here, liquid flows around the outside of the tube while gas flows in the central core. Further increases in gas velocities lead to entrainment of some of the liquids as droplets carried along the central gas core.

At very high mass flowrates, the pattern is that of dispersed flow. At low qualities this is seen as a froth of tiny bubbles uniformly distributed in the liquid. At high qualities it is seen as a mist of fine droplets suspended in vapor.

Flow pattern configurations in upwardly inclined and vertical flow are shown in Figure 8-2. A significant change in flow patterns occurs when a line is inclined slightly to the horizontal. The stratified flow

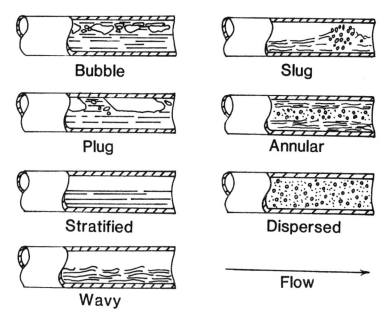

Figure 8-1. Flow pattern configuration in horizontal flow.

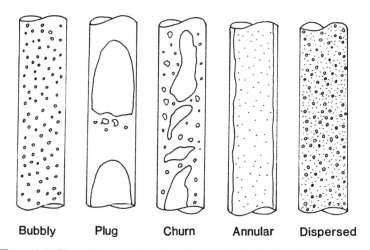

Figure 8-2. Flow pattern configurations in upwardly inclined and vertical flow.

pattern disappears and is replaced by intermittent flow; the criteria for transition from stratified to intermittent flow is[1,2]

$$\sin \theta > D/L \tag{8-1}$$

where θ is the angle of inclination, and L is the length of line being examined.

Wavy flow is still seen in inclined lines, but the region is restricted. As the inclination angle increases, the gas flowrate at which wavy flow begins also increases. At a sufficiently sharp angle of inclination, wavy flow disappears altogether. However, the bubble flow area expands as inclination is increased.

In vertical lines and lines at very sharp angles of inclination, slug flow disappears and the only flow patterns seen are bubbly, plug, churn, and annular. *Churn flow* is a chaotic mixture of large packets of gas and liquid, which seem to have a churning motion.

Flow Pattern Maps

Determination of flow pattern behavior has largely been accomplished by visual observation of the flow in transparent pipes. Experimental flow pattern observations are presented in a flow pattern map. The coordinates of such maps are often superficial mass or linear velocities for the vapor and liquid. A symbol indicating the flow pattern is placed at each location where an observation has been made. Rough transition lines denoting the change from one pattern to another can then be constructed.

One of the most widely used flow maps for horizontal flows is the Baker-Scott plot (Figure 8-3). The map is plotted in terms of G_a/λ versus $G_L\lambda\psi/G_a$, where G_a and G_L are the superficial mass velocities of the liquid and vapor, respectively, and factors ψ and λ are given by

$$\lambda = \left(\left(\frac{\rho_G}{\rho_a} \right) \left(\frac{\rho_L}{\rho_w} \right) \right)^{1/2} \tag{8-2}$$

$$\psi = \left(\frac{\sigma_w}{\sigma} \right) \left(\left(\frac{\mu_L}{\mu_w} \right) \left(\frac{\rho_w}{\rho_L} \right)^2 \right)^{1/3} \tag{8-3}$$

where μ is viscosity, ρ is density, and σ is surface tension. Subscripts a and w refer to physical properties of air and water at atmospheric conditions; subscripts L and G indicate properties of the actual liquid and

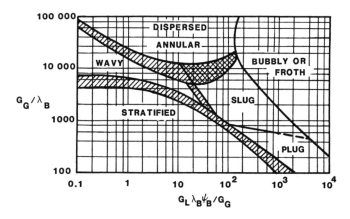

Figure 8-3. Baker-Scott flow map.

vapor flowing in the tube. The plot provides good predictions of flow patterns for line sizes under 2-in., but poor predictions for larger sizes with fluids other than air-water.

For larger pipe sizes, the generalized flow map of Weisman and Kang[1] (Figure 8-4), along with the formulas for pipe diameter and property corrections in Table 8-1, can be used. The generalized flow map in Figure 8-4 was developed for adiabatic, horizontal, and near-horizontal flows.

Figure 8-5 shows a map for vertical and inclined lines.[2] Values of ϕ_1 and ϕ_2 to be used are also given in Table 8-1. Note V_{SG}, V_{SL} are the gas and liquid superficial velocities, respectively, based on one phase flowing separately.

Figure 8-6 provides flow pattern transitions in vertically downward flow. This should only be used as a rough guide.

Sample Calculation 8-1. Kerosene (1000 lbs/hr) and methane gas (300 lbs/hr) are flowing concurrently in a 6-in. ID horizontal pipe. Es-

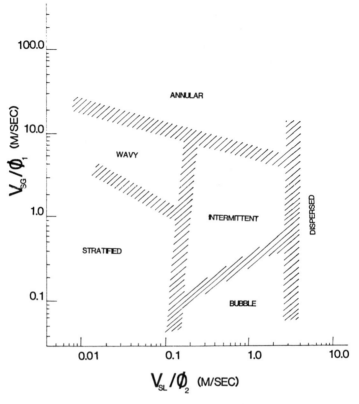

Figure 8-4. Generalized flow map for horizontal flow.[2]

timate the flow regime from the Baker-Scott plot. Physical properties are as follows:

	Density $\rho(\text{lb/ft}^3)$	Viscosity $\mu(\text{lb/h-ft})$	Kinematic Viscosity $\nu(\text{m}^2/\text{s})$	Surface Tension $\sigma(\text{m/h})$
Kerosene	51.2	4.35	2.195×10^{-2}	22
Methane	0.045	2.63×10^{-2}	0.1510	

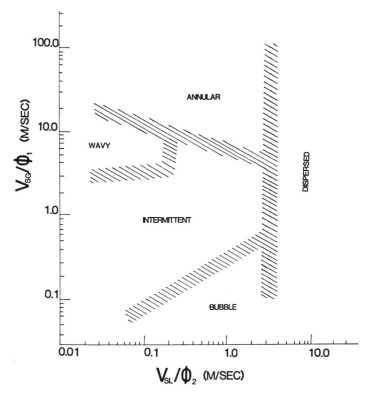

Figure 8-5. Generalized flow map for vertical flow.

Solution

$$\lambda = \left(\left(\frac{\rho_G}{.075}\right)\left(\frac{\rho_L}{62.3}\right)\right)^{1/2} = \left(\left(\frac{0.045}{0.075}\right)\left(\frac{51.2}{62.3}\right)\right)^{1/2} = 0.702$$

$$\psi = (73/\sigma)\left(\mu_L\left(\frac{62.3}{\rho_L}\right)^2\right)^{1/3}$$

$$\psi = \left(\frac{73}{22}\right)\left(4.35\left(\frac{62.3}{51.2}\right)^2\right)^{1/3} = 6.174$$

Table 8-1

Property and Pipe Diameter Corrections to Overall Flow Map

Flow Orientation		ϕ_1	ϕ_2
Horizontal, vertical and inclined flow	Transition-to-dispersed flow	1.0	$\left(\dfrac{\rho_L}{\rho_{sL}}\right)^{-0.33}\left(\dfrac{D}{D_s}\right)^{0.16}\left(\dfrac{\mu_{sL}}{\mu_L}\right)^{0.09}\left(\dfrac{\sigma}{\sigma_s}\right)^{0.24}$
	Transition-to-annular flow	$\left(\dfrac{\rho_{sG}}{\rho G}\right)^{0.23}\left(\dfrac{\Delta\rho}{\Delta\rho_s}\right)^{0.11}\left(\dfrac{\sigma}{\sigma_s}\right)^{0.11}\left(\dfrac{D}{D_s}\right)^{0.415}$	1.0
Horizontal and slightly inclined flow	Intermittent-separated transition		$\left(\dfrac{D}{D_s}\right)^{0.45}$
Horizontal flow	Wavy-stratified transition	$\left(\dfrac{D_s}{D}\right)^{0.17}\left(\dfrac{\mu_{sG}}{\mu_{sG}}\right)^{1.55}\left(\dfrac{\rho_{sG}}{\rho_G}\right)^{1.55}\left(\dfrac{\Delta\rho}{\Delta\rho_s}\right)^{0.69}\left(\dfrac{\sigma_s}{\sigma}\right)^{0.69}$	1.0
Vertical and inclined flow	Bubble-intermittent transition	$\left(\dfrac{D}{D_s}\right)^{n}(1 - 0.65\cos\theta)$ $n = 0.26 e^{0.17(V'_{sL}/V'_{sL})}$	
Inclined flow	Wavy-intermittent transition	$1 + 5\sin\theta$	1.0

ᵃs denotes standard conditions. $D_s = 1.0$ in. $= 2.54$ cm, $\rho_{sG} = 0.0013$ kg/l, $\rho_{sL} = 1.0$ kg/l, $\mu_{sL} = 1$ cP, $\sigma_s = 70$ dynes/cm, $V'_{sL} = 1.0$ ft/sec $= 0.305$ m/sec.

All Air-Water Data

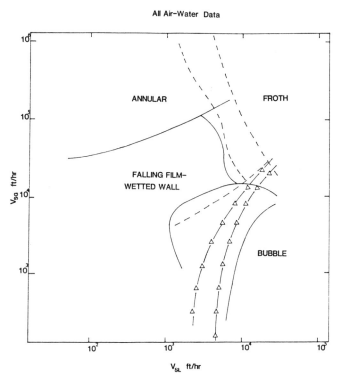

Figure 8-6. Flow pattern transitions for vertically downward flow.[2]

Calculate the superficial mass velocities:

$$A = \frac{1}{4}\pi D^2 = 0.1963 \text{ ft}^2$$

$$G_G = 300/0.1963 = 1.528 \times 10^3 \text{ lb/h-ft}^2$$

$$G_L = 10^3/0.1963 = 5.094 \times 10^3 \text{ lb/h-ft}^2$$

The ordinate:

$$G_G/\lambda = 1.528 \times 10^3/0.702 = 2.177 \times 10^3 \frac{\text{lb}}{\text{h-ft}^2}$$

The abscissa:

$$\frac{G_L \lambda \psi}{G_G} = \frac{(5.094 \times 10^3)\,(0.702)\,(6.174)}{1.528 \times 10^3} = 14.45$$

This point lies in the stratified region of the Baker-Scott plot (Figure 8-3).

Pressure Drop and Holdup

Frictional pressure drop estimates can be made using the Lockhart-Martinelli correlation. The correlation[3,4] employs information on single-phase flow to correlate data in terms of the following parameters:

$$\phi_{G \text{ or } L} = \left(\frac{(dP_F/dz)}{(dP_F/dz)_{G \text{ or } L}}\right)^{1/2} \tag{8-4}$$

$$\chi = \left(\frac{(dP_F/dz)_L}{(dP_F/dz)_G}\right)^{1/2} \tag{8-5}$$

where $(dP_F/dz)_G$ and $(dP_F/dz)_L$ are the pressure gradients for the gas and liquid phases, respectively, flowing alone in single-phase flow through the line or equipment. The correlation for estimating both pressure drop and holdup is given in Figure 8-7. The correlation distinguishes between laminar (or viscous) single-phase flow and turbulent single-phase flow for each phase. For example, ϕ_{Gvt} refers to ϕ_G calculated for laminar liquid phase flow and a turbulent gas flow. Gas void fraction (holdup) data are also shown in Figure 8-7.

For smooth tubes and for turbulent liquid-turbulent gas, χ_{tt} can be computed from

$$\chi_{tt} = \left(\frac{\mu_L}{\mu_G}\right)^{0.1} \left(\frac{\rho_G}{\rho_L}\right)^{0.5} \left(\frac{1-x}{x}\right)^{0.9} \tag{8-6}$$

where x is the vapor quality (mass fraction of vapor).

The Lockhart-Martinelli correlation gives poor predictions for pipe sizes greater than 3-in. at high pressures (overpredicts by 60%) and overpredicts void fraction by as much as 100% for stratified flows. For the latter, the methods outlined by Cheremisinoff[5] should be consulted.

Note that in estimating mass fluxes, the following definitions should be used:

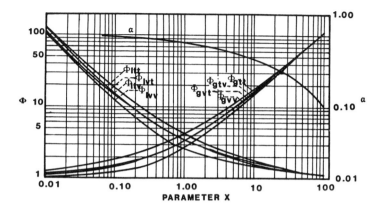

Figure 8-7. Lockhart-Martinelli correlation for frictional pressure drop and holdup.

Liquid

$$G_L = \rho_L u_L (1 - \alpha_0) \tag{8-7}$$

Gas

$$G_G = \rho_G u_G \alpha_0 \tag{8-8}$$

Two-phase density

$$\rho = \alpha \rho_G + (1 - \alpha_0) \rho_L \tag{8-9}$$

where $u_{L,G}$ are local or average velocities of each phase.

Defining G as the total mass flow rate of both phases and x as the mass fraction of gas phase, the following terms are defined:

$$G_L = \frac{G(1 - x)}{(1 - \alpha_0)}, \ u_L = \frac{G(1 - x)}{L(1 - \alpha_0)}$$

$$\tag{8-10}$$

$$G_G = \frac{Gx}{\alpha_0}, \ u_a = \frac{Gx}{\rho_a \alpha_0}$$

Mixture Density and Velocity Ratios

The following formulas provide estimates of two-phase density. The velocity ratio is defined as:

$$K_r = u_G/u_L \tag{8-11}$$

The mixture specific volume is defined as:

$$V_m = \frac{xV_G + K_r(1 - x)V_L}{x + K_r(1 - x)} \tag{8-12}$$

The homogeneous specific volume is that in which $K_r = 1.0$:

$$V_H = V_L\left(1 + \left(\frac{V_G}{V_L} - 1\right)x\right) \tag{8-13}$$

Over a wide range of conditions, the following formulas can be used for the velocity ratio[6]:

$$x > 1, \ K_0 = \left(\frac{V_H}{V_L}\right)^{1/2} \tag{8-14}$$

$$x \leq 1, \ K_0 = \left(\frac{V_G}{V_L}\right)^{1/4} \tag{8-15}$$

References

1. Weisman, J. and Kang, S. Y., *Int. J. Multiphase Flow* (1981).
2. Weisman, J., "Two Phase Flow Patterns," Chapter 15, pp. 409–425, *Handbook of Fluids in Motion,* N. P. Cheremisinoff, et al. (editors), Ann Arbor Science Pub., Ann Arbor, MI (1983).
3. Lockhart, R. W. and Martinelli, R. C., *Chem. Eng. Progress,* 45, No. 1, 39 (1949).
4. Martinelli, R. C. and Nelson, D. B., Trans. ASME, 695, (Aug. 1948).
5. Cheremisinoff, N. P. and Davis, E. J., *AIChE J.,* 25(1), (1979).
6. Chisholm, D., "Gas-Liquid Flow in Pipeline Systems," Chapter 18, pp. 483–513, *Handbook of Fluids in Motion,* N. P. Cheremisinoff, et al. (editors), Ann Arbor Science Pub., Ann Arbor, MI (1983).

9

GAS-SOLID
FLOWS

Flow Through Fixed Beds

The pressure drop across a fixed bed of solids is a function of the average fluid velocity $\bar{v}$, the fluid's viscosity μ_f and density ρ_f, a characteristic dimension representative of the interparticle distance of separation through which the fluid flows d_e, and the bed depth ℓ.

For a randomly packed column,

$$d_e = 4 \times \frac{\text{mean cross section of flow channels}}{\text{mean wetted perimeter of flow channels}} \qquad (9\text{-}1)$$

or

$$d_e = \frac{\text{total bed volume}}{\text{total bed surface}} = \frac{\epsilon}{S} \qquad (9\text{-}2)$$

where: S = surface of solids per unit bed volume
      ϵ = average void fraction

For spherical particles, the equivalent diameter and true surface per unit bed volume are:

$$d_e = \epsilon d_p / 6(1 - \epsilon) \qquad (9\text{-}3)$$

$$S = 6(1 - \epsilon)/d_p \qquad (9\text{-}4)$$

where d_p is the mean particle diameter.

The frictional pressure drop for flow through a fixed bed can be estimated from the Carman-Kozeny equation:

$$\frac{\Delta P}{\ell} = \frac{180(1 - \epsilon)^3 \mu_f U}{g \epsilon^3 d_p}$$

$$= \frac{5(1 - \epsilon)^2 \mu_f U}{g \epsilon^3 (V_p/S_p)^2} \tag{9-5}$$

where V_p and S_p are the particle's volume and surface area, respectively.

The equation is applicable over the range $0.26 < \epsilon < 0.89$ for spherical particles ($d_p = 6V_p/S_p$). For nonspherical particles, an appropriate shape factor can be applied to the particle size.

The specific surface of a bed can be estimated from pressure drop measurements by rearranging Equation 9-5:

$$S_p/V_p = \sqrt{\frac{g \Delta P \epsilon^3}{5(1 - \epsilon)^2 \mu_f U \ell}} \tag{9-6}$$

Estimates of the particle specific surface should be considered as minimum, since the equation does not account for blocked pores or particle geometry.

A more accurate estimate of pressure drop across a fixed bed is the Ergun equation:

$$\frac{\Delta P}{\ell} g_c = 150 \frac{(1 - \epsilon_m)^2}{\epsilon_m^3} \frac{\mu_f U}{(\phi_s \overline{d_p})^2} + 1.75 \frac{1 - \epsilon_m}{\epsilon_m^2} \frac{\rho_f U^2}{\phi_s d_p^2} \tag{9-7}$$

ϕ_s is the particle shape factor, and subscript m refers to mean values.

There are two limiting cases for the Ergun equation; namely, at low Reynolds numbers (Re), the viscous term dominates:

$$\frac{\Delta P}{\ell} g_c = 150 \frac{(1 - \epsilon_m)^2}{\epsilon_m^3} \frac{\mu_f U}{(\phi_s d_p)^2}; \ Re_p < 20 \tag{9-8}$$

At high Re values, the kinetic energy term prevails:

$$\frac{\Delta P}{\ell} g_c = 1.75 \frac{(1 - \epsilon_m)}{\epsilon_m^3} \frac{\rho_f U^2}{\phi_s \overline{d_p}}; \ Re_p > 1000 \tag{9-9}$$

where:

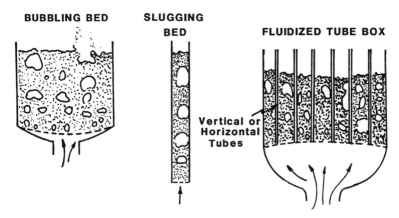

Figure 9-1. Three types of fluid beds.

$$Re_p = \frac{\bar{d}_p \rho_f U}{\mu_f} \tag{9-10}$$

Onset of Fluidization

Loose solids fluidize in various ways: as spouted beds, churning beds, teeter beds, etc. Some of these are useful for drying operations, combustion and incineration, and for various noncatalytic reactor systems. For catalytic reactions, the systems illustrated in Figure 9-1 are best suited; namely, bubbling beds, slugging beds, and fluidized tube boxes. In a fluidized bed the vessel is much larger than the bubble size; hence, bubbles coalesce and grow as they rise and are free from wall effects. In a slugging bed the vessel diameter is small and bubbles quickly grow to the vessel diameter. In a fluidized tube box, bubbles do not rise unimpeded but are influenced by walls and tubes. The behavior is somewhere between slugging and freely bubbling beds.

The key to design and scale-up is to properly characterize the gas flow patterns and contacting between gas and solids.

The onset of fluidization in a bed of loose solids occurs when the drag force exerted by the upward moving gas matches the weight of particles in the bed:

$$\Delta PF_b = W = (F_b \ell_{mf})(1 - \epsilon_{mf})(\rho_s - \rho_f) \qquad (9\text{-}11)$$

where: F_b = cross-sectional area of the bed of solids
ℓ_{mf} = expanded bed height at minimum fluidization
ϵ_{mf} = bed voidage at minimum fluidization conditions
The pressure drop occurring at minimum fluidization is

$$\frac{\Delta P}{\ell_{mf}} = (1 - \epsilon_{mf})(\rho_s - \rho_g) \qquad (9\text{-}12)$$

At the onset of fluidization, the voidage (referred to as the *minimum gas voidage*) is slightly higher than in a packed bed. It essentially corresponds to the loosest state of a packed layer of solids having a small weight. As such, a value for ϵ_{mf} can be estimated from random packing data. The minimum gas voidage can be measured by subjecting a small column of a known quantity of solids to a rising gas stream which initiates incipient particle motion. By recording the expansion of the bed and from knowledge of the buoyant weight of the bed W, the column's cross section F, and solids and gas densities:

$$\epsilon_{mf} = 1 - \frac{W}{\ell_{mf} F(\rho_s - \rho_g)} \qquad (9\text{-}13)$$

Geldart[1] has classified solids into four general classes on the basis of particle size and density relative to the fluid medium. Group A solids are those materials where fluidization is of the bubbling type (aggretative, heterogeneous), although these powders may exhibit particulate fluidization (homogeneous, nonbubbling) over a limited range of velocities near the minimum fluidization point. With these materials, the bed uniformly expands as the gas velocity is increased, reaches a maximum height at the velocity where the bubbles first appear, and gradually collapses to a minimum height with further increases in velocity. Beyond this point, bubbly flow exists, and the bed again expands with increasing velocity. A typical example of a material falling in Group A is commercial silica-alumina cracking catalyst. Solids that are coarser and more dense have different fluidizing characteristics. In this case, bubbles begin to form at the minimum fluidization point and the bed expands in a nonuniform manner. These materials are classified as Group B solids. Group C solids consist of powders which are too fine for normal fluidization because of their cohesive properties. Finally, Group D is comprised of very large and/or dense particles which can produce spouted beds. Figure 9-2 shows Geldart's classification diagram.[2]

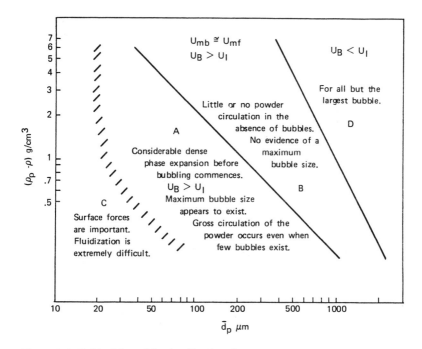

Figure 9-2. Geldart's[1] particle classification diagram.

Minimum Fluidization and Bubbling Velocities

An approximate method for estimating the minimum fluidization velocity has as its basis the Ergun equation. The analysis is outlined by Cheremisinoff[2] and the functional form is as follows:

$$Re_{mf} = \frac{d_p U_{mf} \rho_f}{\mu_f} ((33.7)^2 + 0.0408 \ Ar)^{1/2} - 33.7 \qquad (9\text{-}14)$$

where Ar is the Archimedes number and

$$Ar = \frac{\bar{d}_p^{\,3} \rho_f (\rho_s - \rho_f) g}{\mu_f^2} \qquad (9\text{-}15)$$

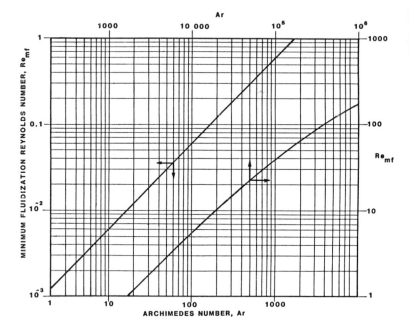

Figure 9-3. Plot of approximate form of extrapolated Ergun equation for estimating the minimum fluidization velocity.

and where

$$Re_{mf} = \frac{\bar{d}_p \rho_f U_{mf}}{\mu_f}$$

The approximate form is shown plotted in Figure 9-3.

For a gas with properties similar to air, the following semiempirical correlation can be used to estimate minimum fluidization velocity for Ar < 40,000:

$$U_{mf} = 0.0008 \ g(\rho_p - \rho)d_{p_{50}}^2/\mu_f \qquad (9\text{-}16)$$

where $d_{p_{50}}$ is the 50 weight % particle size in m, and all other units are in SI.

Minimum bubbling velocity is usually very close to U_{mf} for Group A powders. It is the superficial gas velocity at which the bed first begins to form bubbles. The ratio of U_{mb}/U_{mf} provides a relative index for the range of particulate fluidization

$$\frac{U_{mb}}{U_{mf}} = 0.042\frac{\mu_f^{0.6}\rho^{0.06}}{gd_{vs}^{1.3}(\rho_p - \rho)}$$ (9-17)

d_{vs} is the particle volume surface diameter and all units are in SI. Note that this is an approximate formula developed for silica-alumina catalyst. It has not been widely tested for materials of greatly different properties. An approximation of minimum bubbling velocity can also be obtained from:

$$U_{mb} = 33(\rho_f/\mu_f)^{0.1}d_{p_{50}}$$ (9-18)

Again, units are in SI.

Bed Expansion

For Group A solids (those materials which can be readily fluidized), bed expansion can be expressed in terms of the change in bed height, bed density, or voidage.

Changes in height of the dense phase are observed to be typically about half of that measured for the bed at the bubble point. If the expansion occurs uniformly with increasing gas velocity, then it is due to the increase in bubble holdup:

$$\Delta h/\Delta h_{mf} = \frac{V_B/F}{\bar{U}_B - (V_B/F)} = \frac{U_D - U_i}{\bar{U}_B - (U_D - U_i)}$$ (9-19)

where: V_B = volumetric flow of gas in the form of bubbles
   F = bed's cross section
   $\bar{U}_B$ = mean gas velocity across the distributor
   U_i = superficial gas velocity through the dense phase

The rise velocity depends on the bubble size. For a single rising bubble,

$$\bar{U}_B = 0.7\sqrt{gD_B}$$ (9-20)

For a single rising gas slug,

$$\bar{U}_B = 0.35\sqrt{gD_B}$$

Cheremisinoff et al.[4] gives the following theoretical expression for bed expansion of freely bubbling solids. The expression appears to provide good estimates of bed expansion and fluid bed density over a wide range of conditions, including high pressures

$$\psi = \frac{h}{h_0} = \frac{1}{1 + 2(kFr)^{1/2}} \tag{9-21}$$

where ψ is defined as the relative bed expansion (ratio of expanded to initial or dumped bed height) and Fr is the Froude number, where

$$Fr = \frac{U_s}{gh_0} \tag{9-22}$$

where: h_0 = initial bed height
$\quad\quad\quad U_s$ = gas superficial velocity

Note also that
$$\frac{\rho_{bed}}{\rho_0} = \frac{h_0}{h} = 1 + 2\sqrt{kFr} \tag{9-23}$$

k is a correction factor for the gas density. For air at ambient conditions, $k = 1$. It can be estimated for other properties by the ideal gas law.

Pneumatic Transport

Several flow regimes are observed in both dilute and dense phase conveying. Transition from one regime to another depends on the line orientation, particle and gas properties, gas velocity, and the particle loading concentration. The various regimes are illustrated in Figure 9-4. In dilute-phase vertical transport, when gas velocities are sufficiently high, both phases flow uniformly up the pipe. In this mode the frictional pressure drop is significant. At lower velocities, the solids near the walls of the line decelerate and eventually have no net vertical movement. These particles appear to swirl, float, or flutter near the wall. The solids rate is maintained by flow through the central core of the line, and essentially all of the solids' pressure drop is due to the head of solids in the pipe. At even lower gas velocities, the solids in the vicinity of the wall move downward, while the solids rate is maintained by the upward flow of material in the central region of the pipe. The total pressure drop due to solids can be less than that from the head of solids in the line in this annular-type flow mode. In general, this dilute phase flow regime is unstable and small variations in the gas velocity

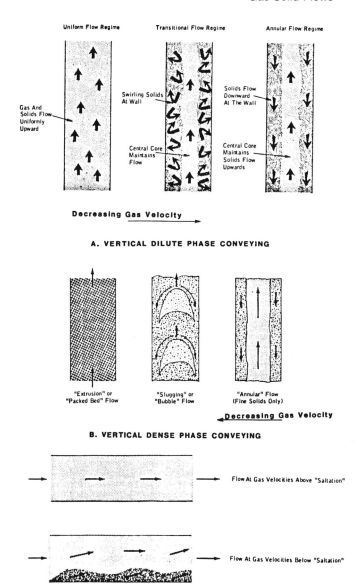

Figure 9-4. Flow regimes in dilute- and dense-phase transport.

can result in large differences in pressure drop. Coarse particles are unable to make the transition to annular flow. These particles tend to exhibit a dense phase type of slugging flow instead of annular flow. Slugging results in significantly higher pressure losses which eventually lead to choked flow conditions (i.e., limited pressure drop is available for conveying).

Different flow regimes are observed in horizontal line orientation in dilute phase transport, in particular, two distinct regimes at gas velocities above and below the saltation velocity. At gas velocities above saltation, the solids flow appears dispersed and greatly resembles the uniform flow regime observed in vertical flows. That is, solids are suspended and move along with the gas and, for the most part, are uniformly distributed across the line. In this regime the pressure drop increases with increasing gas velocity. At lower gas velocities, however, the carrying capacity of the line is eventually exceeded and solids begin to deposit onto the floor. (The gas velocity at which deposition occurs is referred to as the saltation velocity.) At velocities below the saltation velocity, the pressure drop increases with decreasing velocity due to the increased solids deposition. With large particles, saltation phenomena quickly result in a significant rise in pressure drop. For fine solids, the deposition occurs slowly and nonuniformly down the length of the line; consequently, the overall pressure drop increases more slowly but is erratic. This is due to the unstable nature of the nonuniformly sized and distributed deposits which roll down the pipe much like dunes in a desert.

Vertical dense-phase conveying is most commonly employed in stand pipes and risers in petroleum and petrochemical plant operations. In addition to the annular flow regime previously described for fine particles, slugging and "extrusion" or "packed-bed" flow are observed.

For approximate estimates of line sizes and pressure drop for fan or blower sizing, the following method of Branan[5] can be used:

Step 1. Estimate the air velocity required (see Table 9-1 for typical values of different materials). As a rule of thumb, 5000 fpm is appropriate for most materials.

Step 2. Table 9-2 provides typical capacity ranges and line sizes that can be used to obtain initial estimates.

Step 3. Estimate pressure losses: these include material losses (acceleration losses E_1, lifting energy E_2, horizontal losses E_3, losses in bends and elbows E_4) and air losses.

Step 4. Evaluate material losses in ft-lb/min:

Table 9-1
Material Characteristics of Pneumatic Conveying

Material	Specific Wt. (lb/ft³)	Bulk Wt. (lb/ft³)	Abrasiveness	Material	Specific Wt. (lb/ft³)	Bulk Wt. (lb/ft³)	Abrasiveness
Alfalfa meal	35	17	N	Cement, portland	100	80	VA
Almonds, broken or whole	60	29	N	Cement, clinker	131	78	VA
Alum	103	55	N	Chalk, crushed	145	87	A
Alumina	250	60	VA	Chalk, pulverized, minus 100 mesh	143	73	A
Aluminum	165	160	M	Charcoal	25	21	N
Ammonium chloride, crystalline	94*	52	M	Cinders, coal	46	43	A
Antimony	414	417	VA	Clay, dry	85*	63*	A
Apple pomace, dry	30	15	N	Clover seed	75	48	N
Asbestos, shred	153	23	M	Coal, bituminous	87*	50	A
Ashes, Hard Coal	31	35	VA	Coal, anthracite	100*	55	A
Ashes, Soft Coal	43	43	VA	Coca, powdered	70*	35	N
Asphalt, crushed	87	45	VA	Coca beans	80*	40	N
Ammonium sulfate	113	45	M	Coconut, shredded	45*	22	N
Bagasse	20	8	M	Coffee	48	25	N
Bakelite, powdered	100*	40	N	Coke, bituminous	83	30	A
Baking powder	80	41	N	Coke, petroleum	110	40	A
Barley	60	38	N	Copper	556	552	VA
Bauxite, crushed	158	80	VA	Copra (dried coconut)	45	22	M
Beans, meal etc.	82*	41	N	Cork, fine ground	30	15	M
Bentonite	110*	51	VA	Corn, cracked, shelled etc.	70	50	N
Bicarbonate of Soda	137*	41	N	Cornmeal	80	40	N
Bonemeal	75	55	M	Cottonseed	80	40	N
Bones, crushed, minus 1/2"	80	38	M	Cullet (broken glass)	140*	100	A
Bones, granulated or ground, minus 1/8"	100*	50	M	Dicalcium phosphate	144*	43	M
Boneblack	65	23	M	Dolomite	181	100	A
Bonechar	80	40	M	Ebonite, crushed	72	59	N
Borax, powdered	109	53	VA	Egg powder	35*	16	N
Bran	35	16	N	Epsom salts	162*	45	M
Brass	530	165	M	Feldspar	160	70	A
Brewers grain, spent, dry	65	28	N	Ferrous sulphate	118	60	A
Brick	118	135	A	Fish meal	80*	40*	N
Buckwheat	60	40	N	Flaxseed	70	45	N
Calcium Carbide	137	80	A	Flour	50	35	N
Calcium Carbonate	169	147	A	Flue dust, dry	235	117	M
Carbon, amorphous, graphitic	260	130	M	Fluorspar	200	82	A
Carbon black, pelletized	50	25	M	Fly ash	85	40	VA
Carbon black powder, channel	15	5	M	Fullers earth	95*	47	A
Carbon black powder, furnace	15	5	M	Gelatine, granulated	65*	32	N
Carborundum	250	195	VA	Glass batch	162*	95	A
Casein	80	36	M	Glue, ground	80	40	M
Cast Iron, borings	450	165	VA	Gluten meal	80*	40	N
Cast Iron	450	200	VA	Grains, distillery, dry	70*	30	N
Caustic Soda	88	40*	M	Graphite	132	40	A
Cellulose	94	80*	M	Grass seed	25*	11	A

CODE – VA – very abrasive   A – abrasive   M – mildly abrasive   N – less abrasive

*Estimated

Table 9-1 continued.

Material	Specific Wt. (lb/ft³)	Bulk Wt. (lb/ft³)	Abrasiveness	Material	Specific Wt. (lb/ft³)	Bulk Wt. (lb/ft³)	Abrasiveness
Gypsum	145	75	A	Quartz	165	100	VA
Hops, dry	70*	35	N	Resin	67	35	M
Ice, crushed	57*	40	N	Rice	54	45	N
Llmenite ore	312*	140	VA	Rubber, ground	72	23	N
Iron Cast	450	450	VA	Rubber, hard	74	59	N
Iron oxide	330	100*	VA	Rubber, soft	69	55	N
Lead	710	710	A	Rye	60	45	N
Lead arsenate	400	72*	A	Salt, rock	136	45	A
Lead oxide	567	180*	A	Salt, dry, course	138	50	A
Lignite	85	50*	A	Salt, dry, pulverized	140*	75	A
Lime, ground	87	60	VA	Saltpeter	138	80	N
Lime, hydrated	81	40	A	Sand	150	100	VA
Limestone	163	85	VA	Sandstone	144	95	VA
Lithrage	560	180*	A	Sawdust	35	12	N
Lucite	14	7*	A	Shale, crushed	175	87	A
Magnesite	187	187*	VA	Slag, furnace, granulated	132	62	VA
Magnesium	109	100*	VA	Slate	172	85	A
Magnesium chloride	138	33	A	Soap, chips, flakes	30	15	N
Malt, dry	44	26	N	Soap powder	50	25	N
Manganese sulphate	125	70	A	Soapstone talc	175	62	M
Maple, hard	47	43	N	Soap ash, light	74	35	M
Marble	168	96	A	Soda ash, heavy	134	65	M
Marl	120	80	A	Sodium Nitrate	134	70	A
Mica, ground	175	15	M	Sodium Phosphate	94	45	A
Milk, dried, malted, powdered	70	35	N	Soybeans, meal and whole	90	45	N
Monel metal	554	550	M	Starch	96	40	N
Muriate of potash	160	77	M	Steel	487	100	A
Mustard seed	90	45	N	Steel chips, crushed	487	60	A
Naphthalene flakes	71	45	N	Sugar	105	53	N
Nickel	547	537*	VA	Sugar beet pulp, dry	30	13	N
Oats	40	26	N	Sulphur	126	60	N
Oak	47	15	N	Talc	169	60	M
Orange peel, dry	30	15	N	Tanbark, ground	110	55	M
Oxalic acid crystals	104	60	N	Timothy seed	80	36	N
Peanuts	80	40	N	Tin	457	459	A
Peas, dried	95*	47	N	Titanium	280	100	VA
Peas	75	50	N	Tobacco	50	25	N
Phosphate rock	160	80	VA	Vermiculite ore	160	80	A
Phosphate sand	190	95	VA	Wheat	75	45	N
Pine	27	10	N	White lead	120	74	A
Porcelain	150	75	M	Zinc Oxide	360	35	A

CODE — VA — very abrasive   A — abrasive   M — midly abrasive   N — less abrasive

*Estimated

Table 9-2
Typical Capacity Ranges for Pneumatic Transport Lines[5]

Duct Size (in.)	Flow scfm at 5000 fpm	Friction Losses (in. H_2O/100 ft)	Capacity (Thousands lbs/hr) Negative	Positive
4	440	11.0	2–6	12–40
5	680	8.0	3–10	15–60
6	980	6.3	4–15	20–80
8	1800	4.5	15–30	30–160

$E_1 = MU^2/2g = 108M$ @ 5000 fpm

$E_2 = M(H)$

$E_3 = M(L)(F)$

$E_4 = MU^2/gR(L)(F)(N) = 342(M)(F)(N)$ for 48 in. radius, 90° ell

where: M = solids conveyed, lb/min
   U = gas velocity, fpm
   2g = 2.32×10^5ft/min^2
   H = vertical lift, ft
   L = duct horizontal length, ft
   R = 90° ell radius, ft
   F = coefficient of friction and tangent of solids angle of slide or angle of repose
   N = number of 90° ells. For 45°, 30°, etc., express as equivalent 90° ells by direct ratio (e.g., a 30° ell is 0.33 of a 90° ell)

Step 5. From Table 9-2 and the solids rate, obtain a duct size and flow capacity in SCFM.
Step 6. Compute the material losses in inches of water:

$$\frac{\text{ft-lb/min}}{\text{ft}^3\text{/min} \times 5.2} = \text{in.-}H_2O \tag{9-24}$$

Step 7. Compute the air losses:
   a. Calculate the equivalent length of straight pipe by adding to actual length of straight pipe an allowance for conveying type 90° ells of 1 ft of pipe/in. of diameter (e.g., 4 in.-90° ell = 4 ft of pipe).
   b. Assume the following losses for other items:

Item	In.-H_2O
Duct entry loss	1.9
Y-branch	0.3
Cyclone	3.0
Collector vessel	3.0
Filter	6.0

Step 8. Total the material and air losses.
Step 9. Compute fan or blower horsepower.

Miscellaneous Data and Properties of Powders

Ideal loose solids have no forces of attraction between them. The properties of loose materials—in contrast to fluids and composite solids—are characterized by several parameters which must be measured or known prior to the design of any fluid-solids system.

The *bulk density* of solids is the overall density of the loose material including the interparticle distances of separation. It is defined as the overall mass of the material per unit volume. A material's bulk density is sensitive to the particle size, the mean particle density, moisture content, and the interparticle separation (i.e., degree of solids packing). It is measured simply by pouring a weighed sample of particles through a funnel into a graduated cylinder, from whence the volume occupied determines the loose bulk density. By gently vibrating the container walls, the interparticle distance between particles decreases and hence, the volume decreases. The material thus becomes denser with time, and its bulk density achieves some limiting value ρ_{max} known as the *tapped* or *packed bulk density.*

The ratio of ρ_{max}/ρ_{min} can be as high as 1.52 depending on the material. Consequently, when bulk densities are reported, it is important to note whether the value was determined under loose or tapped conditions, along with the mean particle size. Most literature values report an average bulk density which is representative of the material most often handled. Loose solids may be broadly characterized according to their bulk densities:

Light material $\rho_b < 600$ kg/m^3

Average $600 < \rho_b < 1100$ kg/m^3

Extra heavy $\rho_b > 2000$ kg/m^3

The loose bulk density can be computed as:

$$\rho_b = \frac{G_1 - G}{V}, \text{ kg/m}^3 \qquad (9\text{-}25)$$

where: G_1, G = weights of filled and empty cylinders
$\quad\quad\quad V$ = internal volume of cylinder

Bulk density is related to particle density through the interparticle void fraction ϵ in the sample

$$\rho_b = \rho_p(1-\epsilon) \qquad (9\text{-}26)$$

The value of ϵ varies between the limits of 0 and unity; however, many particles of interest in fluidization have a loosely poured voidage of approximately 0.4–0.45.

Particle density ρ_p (also called apparent density) is the density of a particle including the pores or voids within the individual solids. It is defined as the weight of the particle divided by the volume occupied by the entire particle.

The *skeletal density* ρ_s (also called the *true density*) is defined as the density of a single particle excluding the pores. That is, it is the density of the *skeleton* of the particle if the particle is porous. For nonporous materials, skeletal and particle densities are equivalent. For porous particles, skeletal densities are higher than the particle density.

Measurements of the skeletal density are made by liquid or gas pycnometers. When liquids are used, the pycnometer has a fixed and known volume. A specified weight of solids is immersed in a liquid of known density which wets the solids and penetrates into the pores of the particles. The volume of liquid displaced by the solids is then determined by difference, and the skeletal density of the weighed solids is computed from the measured displaced volume.

Particle and skeletal densities are related through the following equation:

$$\rho_p = \frac{1 + \rho_f \xi}{\dfrac{1}{\rho_s} + \xi} \qquad (9\text{-}27)$$

where: ρ_p = particle density
$\quad\quad\quad \rho_s$ = skeletal density
$\quad\quad\quad \rho_f$ = density of fluid contained in the pores of the solid
$\quad\quad\quad \xi$ = pore volume per unit mass of solids

When the particle pores are saturated with gas, the term $\xi\rho_f$ is negligible and Equation 9-27 reduces to

$$\frac{1}{\rho_p} = \frac{1}{\rho_s} + \xi \tag{9-28}$$

The pore volume can be measured indirectly from the adsorption and/or desorption isotherms of equilibrium quantities of gas absorbed or desorbed over a range of relative pressures. Pore volume can also be measured by mercury intrusion techniques, whereby a hydrostatic pressure is used to force mercury into the pores to generate a plot of penetration volume versus pressure. Since the size of the pore openings is related to the pressure, mercury intrusion techniques provide information on the pore size distribution and the total pore volume.

Moisture can significantly affect loose materials, particularly their flowability. Low temperatures, particle bridging, and caking can alter interparticle void fractions and cause dramatic changes in bulk density. The terms *moisture* or *moisture content* are used to denote the degree of liquid retained on or in solids.

Moisture is defined as the ratio of the fluid's weight retained by solids to the weight of wet material:

$$W = \frac{G_w - G_d}{G_w} \tag{9-29}$$

where G_w and G_d are the weights of wet and absolute dry material, respectively. Moisture content W_c is the ratio of the moisture weight to the weight of absolute dry material:

$$W_c = \frac{G_w - G_d}{G_d} \tag{9-30}$$

Values of W and W_c can be expressed as either fractions or percents. The presence of moisture tends to increase the relationship between moisture content and the density of loose or lump materials as follows:

$$\rho_m = \rho(1 + W_c) \tag{9-31}$$

And, for dusty and powdered materials,

$$\rho_m = \rho\frac{1 + W_c}{\left(1 + \frac{W_c}{3}\rho_p/\rho_f\right)} \tag{9-32}$$

where: ρ_m, ρ = densities of wet and dry loose materials
  ρ_p = particle density
  ρ_f = density of liquid filling the solid particle pores

In addition to the physical properties described above, there are those properties which affect the flowability of the material. Specifically, these properties are the material's angle of repose, angle of internal friction, and the angle of slide.

The *angle of repose* is defined as the angle between a line of repose of loose material and a horizontal plane. Its value depends on the magnitude of friction and adhesion between particles and determines the mobility of loose solids which is a critical parameter in the design of conical discharge and feeding nozzles and in establishing vessel geometries. In all cases, the slopes of such nozzles should exceed the angle of repose.

The angle of repose is the measured angle between a horizontal plane and the top of a pile of solids. The poured angle of repose is obtained when a pile of solids is formed, whereas the *drained angle* results when solids are drained from a bin. Figure 9-5 distinguishes between the two terms signifying the angle of repose. For monosized particles or particles with a relatively narrow size distribution, the drained and poured angles are approximately the same. If, however, the solids have a wide size distribution, the drained angle is higher than the poured angle.

In bin design the drained angle is more important. However, the differences between the angles are rarely important and, since the poured angle is easier to measure, it is most frequently reported in the literature. In general, the lower the angle of repose, the more free flowing the material, and hence, the shallower the bin angle required. Materials can be roughly categorized according to their angle of repose as follows:

For very free-flowing granules:	$25° < \beta < 30°$
For free-flowing granules:	$30° < \beta < 38°$
For fair-to-passable flow of powders:	$38° < \beta < 45°$
For cohesive powders:	$45° < \beta < 55°$
For very cohesive powders:	$55° < \beta < 70°$

The angle of repose is sensitive to the condition of the supporting surface. The smoother the surface, the smaller the angle. The angle may also be reduced by vibrating the supporting surface. When handling slow-moving materials with large angles of repose, well-designed bun-

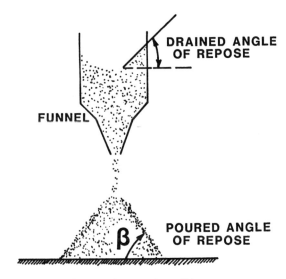

Figure 9-5. The angle of repose for granular solids.

kers and hoppers are provided with highly polished internal surfaces and low-amplitude vibrators.

The *angle of internal friction* α is defined as the equilibrium angle between flowing particles and bulk or stationary solids in a bin. Figure 9-6 illustrates the definition. The angle of internal friction is invariably greater than the angle of repose.

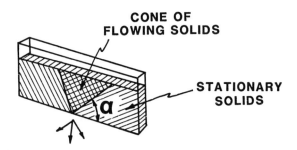

Figure 9-6. The angle of internal friction.

The *angle of slide* is defined as the angle from the horizontal of an inclined surface on which an amount of material will slide downward due to the influence of gravity. It is an important consideration in the design of chutes and hoppers as well as pneumatic conveying systems. The angle of slide provides a measure of the relative adhesiveness of a dry material to a dissimilar surface. The angle depends on the type of solids, the nature and surface properties (e.g., roughness and dryness), the surface configuration (e.g., degree of curvature), the manner in which the solids are placed on the surface, and the rate of change of the slope of the surface during measurements. Table 9-3 provides typical values of loose solids as measured without vibration of the horizontal support surface.

The term *particle size* requires careful definition. Diameters of irregular particles may be defined in terms of the geometry of the individual

Table 9-3
Properties of Loose Solids

Material	Bulk Density (g/cm^3)	Angle of Repose (deg)	Friction Coefficient Inside	Friction Coefficient Outside on Steel
Sulfur	0.67	40	0.8	0.625
Magnesite—Caustic	1.02	31	0.575	0.5
Magnesium Oxide	0.47	36	0.49	0.37
Phosphate Powder	1.52	29	0.52	0.48
Calcium Chloride	0.68	35	0.63	0.58
Napthalene—Crushed	0.57	37	0.725	0.6
Anhydrous Sodium Carbonate	0.585	41	0.875	0.675
Sodium Chloride—Fine	0.93	38	0.725	0.625
Carbamide—Powdered	0.54	42	0.825	0.56
Superphosphate—				
Granulated	1.1	31	0.64	0.46
Powdered	0.8	36	0.71	0.7
Silicylic Acid—Powdered	0.46	44	0.95	0.78
Talc	0.85	40	—	—
Cement	1.15	30	0.5	0.45
Chalk	1.1	42	0.81	0.76
Sand—Fine	1.51	33	1.0	0.58
Graphite	0.45	40	—	—
Coal—Fine	0.95	36	0.67	0.47
Earth—Dry	1.2	30	0.9	0.57
Wheat	0.77	29	0.35	0.28
Pea	0.743	26	0.4	0.42

particulates. These diameters are usually referred to as statistical diameters since large numbers of particles must be measured and averaged to provide a representative size.

One definition for particle diameter is based on its linear dimensions and is characterized as a mean arithmetic value:

$$d_i = \frac{\ell bh}{3} \tag{9-33}$$

Particle size distributions can be conveniently expressed as a cumulative distribution, i.e., in terms of particle size versus weight, volume or number fraction of particles smaller or larger than the stated particle size. Graphically, cummulative particle size distributions can be represented on log-normal probability, normal probability and Rosin-Rammler plots.

In a *log-normal probability distribution* the logarithm of the particle sizes is normally distributed. A cumulative log-normal distribution will result in a straight line on "log-probability" coordinates. The standard deviation for this distribution is defined as:

$$\sigma_s = \frac{50\% \ \text{particle size}}{15.9\% \ \text{particle size}} \tag{9-34}$$

The greater the value of σ_s, the wider the particle size distribution.

A *normal probability* or Gaussian distribution is essentially a standard definition in which a plot of frequency versus particle size results in a symmetrical bell-shaped curve. The equation of the ordinate is:

$$Y = \frac{1}{\sqrt{2\pi}\sigma_s} \exp\left[-\frac{1}{2}\left(\frac{\alpha' - \bar{\alpha}'}{\sigma_s}\right)^2\right] \tag{9-35}$$

And, for the abscissa,

$$X = \frac{\alpha' - \bar{\alpha}'}{\sigma_s} \tag{9-36}$$

where $\bar{\alpha}'$ is the average size or parameter obtained by dividing the sum of the individual values by the frequency

$$\bar{\alpha}' = \frac{\Sigma\alpha'}{n} \tag{9-37}$$

and the deviation from the mean is defined as the value minus the average value:

$$\alpha'' = \alpha' - \bar{\alpha}' \tag{9-38}$$

The standard deviation is then equal to the square root of the sum of the deviations squared divided by the frequency

$$\sigma_s = \sqrt{\frac{\Sigma(\alpha'')^2}{n}} \tag{9-39}$$

When plotted on probability coordinates, a cumulative normal distribution gives a straight line. The mean particle size is the 50% point on the probability scale.

Particle size can be described in terms of the Sauter mean; i.e., the volume surface mean particle size

$$\bar{d}_{vs} = \frac{1}{\Sigma W_i/d_i} \tag{9-40}$$

where: W_i = weight fraction of the ith particle size range
  d_i = average particle diameter in the ith particle size range

Another useful characteristic size is a median diameter, denoted as $d_{p_{50}}$. This is the diameter at which 50% of some property, such as weight, number, surface area, etc., of the distribution is due to particles smaller than d_{50}. It can be directly obtained from the 50% point on a cumulative plot of the size distribution. The most frequently used median diameters are the weight median size (i.e., 50% point on a weight % versus size cumulative plot), and the number median diameter. For normal distributions, the mean and median sizes are the same. Common definitions of particle sizes are summarized in Table 9-4.

Particle shape influences the flow characteristics of loose solids. There are a variety of measures reported in the literature of nonsphericity of particles with the following definition used:

$$\phi_s = \frac{F_s}{F_p} \tag{9-41}$$

where F_s and F_p are the surface areas of a sphere and the particle having identical volumes, respectively. Equation 9-41 defines sphericity over the range of $0 < \phi_s < 1$, with the value of unity corresponding to a sphere. Typical literature values are noted in Table 9-5.

Table 9-4
Common Definitions of Mean Diameters

Name	Common Symbol	Definition
Arithmetic mean	$\bar{d}_{10}$	$\dfrac{1}{n}\Sigma d_i f_i$
Geometric mean	$\bar{d}_g$	$(d_1^{f_1}d_2^{f_2}d_3^{f_3}\ldots d_n^{f_n})1/n$
Harmonic mean	$\bar{d}_{ha}$	$\left[\dfrac{1}{n}\Sigma\dfrac{f_i}{d_i}\right]^{-1}$
Mean surface diameter	$\bar{d}_{20}$	$\left(\dfrac{\Sigma f_i d_i^2}{n}\right)^{1/2}$
Mean weight diameter	$\bar{d}_{30}$	$\left(\dfrac{\Sigma f_i d_i^3}{n}\right)^{1/3}$
Linear mean diameter	$\bar{d}_{21}$	$\dfrac{\Sigma f_i d_i^2}{\Sigma f_i d_i}$
Surface mean diameter	$\bar{d}_{32}$	$\dfrac{\Sigma f_i d_i^3}{\Sigma f_i d_i^2}$
Weight mean diameter	$\bar{d}_{43}$	$\dfrac{\Sigma f_i d_i^4}{\Sigma f_i d_i^3}$

Table 9-5
Shape Factors For Common Materials

Material	Shape Factor, ϕ_5
Sand	$0.534 \sim 0.861$
Silica	$0.554 \sim 0.628$
Pulverized coal	0.696
Bituminous coal	0.625
Celite cylinders	0.861
Iron catalyst	0.578

References

1. Geldart, D., *Powder Technology*, 7, 285 (1973).
2. Cheremisinoff, N. P. and P. N. Cheremisinoff, *Hydrodynamics of Gas-Solids Fluidization*, Gulf Publishing Co., Houston, TX (1984).
3. Geldart, D. and Abrahamson, A. R., *Powder Technology*, 19:133–136 (1978).
4. Cheremisinoff, N. P. and Azbel, D., "Mathematical Model of Freely-Bubbling Fluidized Bed," *Proceedings of 4th International Conf. on Math. Model.*, Zurich, Switzerland (Aug. 1983).
5. Branan, C., *The Process Engineer's Pocket Handbook*, Gulf Publishing Co., Houston, TX (1976).

10

PUMP CALCULATIONS

Types of Pumps

The major types of pumps in the process industries are centrifugal, axial, regenerative turbine, reciprocating, metering, and rotary. There are two categories under which these classes are grouped: (1) dynamic pumps, and (2) positive displacement pumps.

Dynamic pumps include centrifugal and axial types. They operate by developing a high liquid velocity which is converted to pressure in a diffusing flow passage. These pumps, in general, are lower in efficiency than the positive displacement types. However, they do operate at relatively high speeds, thus providing high flow rates in relation to the physical size of the pump. Furthermore, they usually have significantly lower maintenance requirements than positive displacement pumps.

Positive displacement pumps operate by forcing a fixed volume of fluid from the inlet pressure section of the pump into the pump's discharge zone. This is performed intermittently with reciprocating pumps. In the case of rotary screw and gear pumps, the action is continuous. This category of pumps operates at lower rotative speeds than dynamic pumps. Positive displacement pumps also tend to be physically larger than equal-capacity dynamic pumps.

Pump Design Parameters

There are four characteristics descriptive of all pumps, namely:

1. *Capacity* $Q(m^3/sec)$. The quantity of liquid discharged per unit time.
2. *Head* $H(m)$. The energy supplied to the liquid per unit weight. H is obtained by dividing the increase in pressure by the liquid specific

weight. This specific energy is determined by the Bernoulli equation. Head can be defined as the height to which 1 kg of discharged liquid can be lifted by the energy supplied by a pump. Therefore, it does not depend on the specific weight $\gamma(kg_f/m^3)$ or density $\rho(kg/m^3)$ of liquid to be pumped.

3. *Power* $\tilde{N}(kg_f\text{-}m/sec)$. The energy consumed by a pump per unit time for supplying liquid energy in the form of pressure. Power is equal to the product of specific energy H and the mass flow rate γQ:

$$\tilde{N} = \gamma QH = \rho gQH \tag{10-1}$$

Effective power $\tilde{N}_e$ is larger than $\tilde{N}$ due to energy losses in a pump. Its relative value is evaluated by the pump efficency η_p:

$$\tilde{N}_e = \frac{\tilde{N}}{\eta_p} = \frac{\rho gQH}{\eta_p} \tag{10-2}$$

4. *Overall efficiency* η. The ratio of useful hydraulic work performed to the actual work input. It characterizes the pump's performance. The value of η reflects the relative power losses in the pump

$$\eta = \eta_v \times \eta_h \times \eta_m \tag{10-3}$$

where η_v is the volumetric efficiency defined as the ratio of liquid actually pumped to that which theoretically should be discharged. That is, it indicates the percentage losses (or slip). In practice, slip should not exceed 5%. η_h is the hydraulic efficiency defined as the ratio of the actual head pumped to the theoretical head

$$\eta_h = \frac{H}{H + \text{hydraulic losses}} \tag{10-4}$$

Hydraulic losses are those losses as head in the suction and discharge sections of a pump. In the suction end, these losses are comprised of velocity head, entrance head, friction head in the suction line, and losses in bends and in suction valves.

Discharge line losses consist of losses in the discharge valves, velocity head, and friction in the discharge piping.

η_m is the mechanical efficiency defining the relation between the indicated pump horsepower and the actual power input from the drive. It characterizes mechanical losses in the pump (e.g., in bearings, stuffing boxes, etc.).

Overall efficiency η depends on the pump design and varies from 50% for small pumps to about 90% for large sizes. The power consumed by a motor (defined as the nominal power of a motor) $\tilde{N}_m$ exceeds brake power by mechanical losses incurred in transmission and in the motor itself. These losses are accounted for by including the efficiencies of the transmission (η_{tr}) and motor (η_m):

$$\tilde{N}_m = \frac{\tilde{N}_e}{\eta_m \times \eta_{tr}} = \frac{\tilde{N}}{\eta_e \times \eta_m \times \eta_{tr}} \tag{10-5}$$

The product of $\eta_e\,\eta_m\,\eta_{tr}$ is the total efficiency of a pump and may be defined as the ratio of hydraulic power to the motor's nominal power:

$$\eta = \frac{\tilde{N}}{\tilde{N}_m} = \eta_\rho \eta_{tr} \eta_{mot} \tag{10-6}$$

The total efficiency of a pump is expressed by the product of these five values:

$$\eta = \eta_v \times \eta_h \times \eta_m \times \eta_{tr} \times \eta_{mot} \tag{10-7}$$

The actual power of a pump motor $\tilde{N}_A$ should be based on the energy required to overcome the fluid's inertia at startup, so as not to overload the unit.

Coefficient β is determined from the size of the motor, typical values of which are given as follows:

$\tilde{N}_A$(kW):	< 1	1–5	5–50	> 50
β:	2–1.5	1.5–1.2	1.2–1.15	1.1

Head and Suction Head

Head H characterizes the additional energy ($\ell = gH$) added to 1 kg of liquid in a pump and can be evaluated from the Bernoulli equation as outlined earlier. The total head of the operating pump may be evaluated from measurements obtained from the pressure p_m and vacuum p_v gages.

Total suction head can be obtained from the measurement h_{sg} on the gage on the pump's suction nozzle (corrected to the pump centerline and converted to feet of liquid) plus the barometer reading in feet of liquid and the velocity head h_{vs} (feet) at the point of gage attachment:

$$H_s = H_{sg} + atm + H_{vs} \qquad (10\text{-}9)$$

When the static pressure at the suction flange is less than atmospheric, the measurement obtained from a vacuum gauge replaces H_{sg} in Equation 10-9 and is assigned a negative sign.

Total discharge head H_d is obtained from a gage (H_{dg}) at the pump's discharge flange (corrected to the pump centerline and converted to feet of liquid) plus the barometer reading and the velocity head H_{vg} at the point of gage attachment:

$$H_d = H_{dg} + atm + H_{vg} \qquad (10\text{-}10)$$

Again, if the discharge gage pressure is below atmospheric, the vacuum-gage measurement replaces H_{dg} with a negative sign. Before installation, it is possible to estimate the total discharge head from the static discharge head H_{sd} and discharge friction H_{fd} as follows:

$$H_d = H_{sd} + H_{fd} \qquad (10\text{-}11)$$

Static suction head H_{ss} is defined as the vertical distance (feet) between the free level of the source of supply and the pump centerline, plus the absolute pressure at this level (converted to feet of liquid). *Total static head* H_{ts} is the difference of discharge and suction static heads.

The suction generated by a pump is derived from a pressure difference in the suction source (P_1) and the pump (P_s), or under the action of the head difference $P_1/\rho g - P_s/\rho g$. Suction height can be determined as follows:

$$H_s = \frac{P_1}{\rho g} - \left(\frac{P_s}{\rho g} + \frac{w_s^2 - w_1^2}{2g} + h_{\ell s} \right) \qquad (10\text{-}12)$$

or

$$H_s = \frac{P_1}{\rho g} - \left(\frac{P_s}{\rho g} + \frac{w_s^2}{2g} + h_{\ell s} \right) \qquad (10\text{-}13)$$

because $w_1 = 0$.

These expressions show that the suction head increases with increasing P_1 and decreases with increasing w_s and $h_{\ell s}$.

If liquid is pumped from an open tank (i.e., $P_1 = P_a$; where P_a corresponds to atmospheric), the suction pressure P_s must exceed pressure P_t (the pressure of the saturated vapor of the liquid at the pumping temperature ($P_s > P_t$)); otherwise, the fluid begins to boil. When the pumped liquid vaporizes, the suction head goes to zero at the limit and flow stops. Consequently,

$$H_s \leq \frac{P_a}{\rho g} - \left(\frac{P_t}{\rho g} + \frac{w_s^2}{2g} + h_{\ell s} \right) \quad (10\text{-}14)$$

Equation 10-14 shows that the suction head is a function of atmospheric pressure, the fluid velocity and density, temperature (and correspondingly, the liquid's vapor pressure), and the hydraulic resistance of the suction piping. When pumping from an open tank, the suction head cannot exceed the head of pumping liquid which corresponds to atmospheric pressure (the value of which depends on the height of the pump installation above a specified datum (normally sea level)).

At temperatures approaching the boiling point of the liquid, the suction head becomes zero. In this situation the pump must be installed below the suction line to provide a back-liquid. This method is also used for pumping high-viscosity liquids. In addition to evaluating the friction head and local resistance losses, inertia losses (for piston pumps) H_i and the effect of cavitation (for centrifugal pumps) h_k must be accounted for in the overall suction head term.

Head losses due to overcoming inertia forces H_i (in piston pumps) may be estimated by an expression which relates the pressure acting on the piston to the inertia force of a liquid column moving in the suction piping:

$$H_i = \frac{6}{5} \times \frac{\ell}{g} \times \frac{f}{f_1} \times \frac{u^2}{r} \quad (10\text{-}15)$$

where: ℓ = height of liquid column in the piping (for pumps with a gas chamber, it is the distance between pump centerline and the liquid level in the chamber)

g = acceleration due to gravity

f and f_1 = cross-sectional areas of the piston and piping, respectively

u = circumferential crank velocity

r = crank radius

Pump Law Expressions

In centrifugal pumps the liquid flows along the surface of the impeller vanes while the tips of the vanes move relative to the casing of the pump. To develop an expression of the virtual head developed by a centrifugal pump, we shall assume the path followed by a volume of liquid as it passes through the pump in relation to a stationary impeller with the fluid at the same relative velocity as an actual rotating impeller.

The specific work per unit weight of liquid is equal to the specific energy obtained by the fluid in the pump:

$$E = \frac{u_2^2 - u_1^2}{2g} \tag{10-16}$$

The head of the pump is equal to the difference of heads between the pump's inlet and outlet:

$$H_T = H_1 - H_2 = \frac{P_2 - P_1}{\rho g} + \frac{C_2^2 - C_1^2}{2g} \tag{10-17}$$

The virtual head of a centrifugal pump is given by the following equation:

$$H_T = \frac{u_2 C_2 \cos \alpha_2 - u_1 C_1 \cos \alpha_1}{g} \tag{10-18}$$

where C_1 and C_2 are the vector components of centrifugal force acting on the fluid (refer to Figure 10-1). This equation represents the theoretical maximum head that could be developed for a specified set of operating conditions. Note that the liquid entering the pump usually moves along the impeller in the radial direction. In this case the angle between the absolute velocity value of the liquid entering the impeller and the tangential velocity is $\alpha_1 = 90°$ (this corresponds to the liquid entering the impeller without any shock). Equation 10-18 simplifies to:

$$H_T = \frac{u_2 C_2 \cos \alpha_2}{g} \tag{10-19}$$

or

$$H_T = \frac{u_2^2}{g} \left(1 - \frac{w_2}{u_2} \cos \beta_2 \right) \tag{10-20}$$

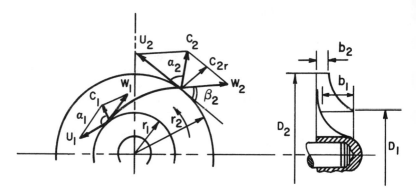

Figure 10-1. Definition of the system for the equations for centrifugal pumps.

From the width of the impeller (b), the length of the circumference $(2\pi r_2)$, and the cross section of the flow leaving the impeller $(2\pi r_2 b)$, the quantity of liquid being pumped is

$$V = 2\pi r_2 b w_2 \sin \beta_2 \qquad (10\text{-}21)$$

Hence,

$$w_2 \cos \beta_2 = \frac{V}{2\pi r_2 \tan \beta_2} \qquad (10\text{-}22)$$

The relationship between the head H and the volumetric rate of flow through the pump V is

$$H = \frac{u_2^2}{g} - \frac{V}{g(2\pi r_2 b) \tan \beta_2} \qquad (10\text{-}23)$$

For a given speed of rotation, there is a linear relation between the head developed and the rate of flow. This is illustrated for different vane configurations in Figure 10-2.

If the outlet vane angles are inclined backward, β_2 is less than 90°; hence, $\tan\beta$ is positive and therefore the head decreases as the throughput increases. If β_2 is greater than 90° (i.e., the outlet vane angles are inclined forward), the head increases at higher throughputs (curve B, Figure 10-2). Radial vanes provide a constant head (curve C,

Figure 10-2). When the flow is zero (V = 0), then, regardless of the vane angle, the head expression is

$$H_T = \frac{u_2^2}{g} \qquad (10\text{-}24)$$

The actual head is always less than virtual head

$$H = H_T \eta_h \epsilon \qquad (10\text{-}25)$$

where η_h is the hydraulic efficiency (with typical values between 0.8 to 0.95), and ϵ is a coefficient which accounts for the number of vanes ($\epsilon = 0.6\text{-}0.8$).

The output and head of a centrifugal pump depend on the number of revolutions per unit of time made by the impeller. Throughput is directly proportional to the radial component of the absolute velocity at the exit from the impeller (i.e., $Q \propto C_{2,r}$). If the number of revolutions is changed from n_1 to n_2 (thus changing from Q_1 to Q_2 correspondingly), the trajectories of the motion of liquid particles remain unaltered and the velocity parallelograms at any corresponding points will be geometrically similar:

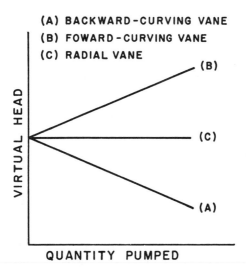

Figure 10-2. Plot of virtual head versus flow capacity for different outlet vane angles.

the velocity parallelograms at any corresponding points will be geometrically similar:

$$\frac{Q_1}{Q_2} = \frac{C_{2r}'}{C_{2r}''} = \frac{u_2'}{u_2''} = \frac{\pi D_2 n_1}{\pi D_2 n_2} = \frac{n_1}{n_2} \qquad (10\text{-}26)$$

The head is proportional to the square of circumferential velocity, i.e.,

$$\frac{H_1}{H_2} = \left(\frac{u_2'}{u_2''}\right)^2 = \left(\frac{n_1}{n_2}\right)^2 \qquad (10\text{-}27)$$

The power developed by the pump is proportional to the product of volumetric flow rate Q and head H:

$$\frac{\tilde{N}_1}{\tilde{N}_2} = \left(\frac{n_1}{n_2}\right)^3 \qquad (10\text{-}28)$$

Equations (10-26) through (10-28) are the *equations of proportionality* or pump law expressions for centrifugal machines. These expressions state the following:

1. A change in the number of impeller revolutions (from n_1 to n_2) causes the pump throughput to change in a directly proportional manner.
2. The heads of the two systems are proportional to the number of revolutions raised to the squared power.
3. The powers developed by pumps are proportional to the number of revolutions to the third power.

Performance Curves

The products of the developed head (in units of pressure) and volumetric flowrate represent the power absorbed by the pumped fluid. Because the head approaches zero at the maximum flow rate, power first increases from zero (i.e., at Q = 0) to a maximum and then decreases to zero at a maximum volumetric flow rate. This can be seen from a pump horsepower versus throughput curve.

Power needed to drive the pump is the same power that is required to overcome all the losses in the system and supply the energy added to the liquid. The losses include frictional losses at the impeller as well as turbulent losses; the disk friction (or energy required to rotate the impeller in the fluid); leakage from the periphery back to the eye of the impeller;

and mechanical friction losses in various pump components such as bearings, stuffing boxes, and wearing rings. The sum of all these power consumption items produces the brake horsepower curve.

Brake horsepower is required even when the volumetric flow rate is zero. With increasing flow rate, the brake horsepower increases even when the head is zero.

An additional crossplot can be obtained by dividing the fluid horsepower by the brake horsepower values (the definition of mechanical efficiency of a pump) and plotting efficiency versus throughput. All three performance curves (head $(H-V)$, power $(\tilde{N}-V)$, and efficiency $(\eta-V)$ are conveniently combined into a single diagram. Figures 10-3 and 10-4 are illustrative only. Manufacturers should always be consulted for the specific performance data of a pump.

Figure 10-3 contains several curves corresponding to different impeller diameters for a specific type of machine. Also shown are several lines of constant brake horsepower and constant efficiency. Such a diagram provides information on the characteristics of a pump for a definite head at a specified liquid flow rate. By specifying coordinates H, V, interpolation between the curves of brake horsepower, impeller di-

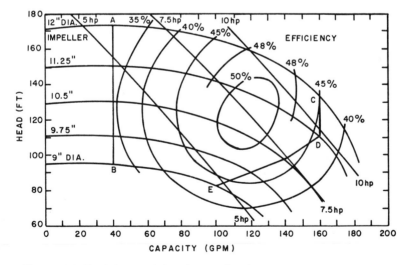

Figure 10-3. Total characteristics of a centrifugal pump.

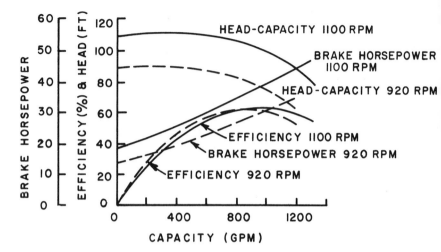

Figure 10-4. Head, efficiency, and brake horsepower plotted as a function of capacity for a centrifugal pump operating at two speeds with the same liquid.

ameter, and efficiency can be obtained on all characteristic values of a pump.

It is important also to evaluate the influence of impeller revolutions on pump performance. An increase in the number of rotations n is accompanied by an increase in head at a constant flow rate.

Pump Selection

In selecting a pump it is necessary to consider all pump system characteristics (i.e., the arrangement of piping, fittings, and equipment through which liquids flow). The characteristics of a pumping system express the relationship between flow rate Q and head H needed for liquid displacement through a given arrangement. Head H is the sum of the geometric height H_g and head losses h. Head losses are proportional to the square of the flow rate:

$$h_\ell = KQ^2 \tag{10-29}$$

where K is a proportionality coefficient. The characteristics of a pump system may be expressed by the following parabolic expression:

$$H = H_g + KQ^2 \tag{10-30}$$

The characteristics of both the pump system and the pump can be represented on a common plot as shown in Figure 10-5. Point A (the intersection of both characteristics curves) represents the *operating point* and corresponds to the maximum capacity of the pump Q while operating for a pump system. If a higher capacity is required, it is necessary either to increase the number of motor rotations or to change to a larger pump. Increased capacity can also be achieved by decreasing the hydraulic resistance of the pump system (h_ℓ). In this case the operating point is displaced along the pump characteristics toward the right. Hence, a pump should be selected so that the operating point corresponds to a desired head and capacity.

Series and Parallel Pumping

In series pumping the pump head component of total available head is the sum of the heads developed by each pump at any given flow. Each pump must be selected to operate satisfactorily at the system design flow. Pumps operating in series are referred to as "pressure additive." This principle is illustrated in Figure 10-6. With one pump operating, the system flow will occur at point A. With both pumps operating, the

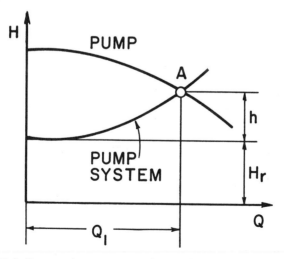

Figure 10-5. Pump and pump system characteristics represented on a single plot.

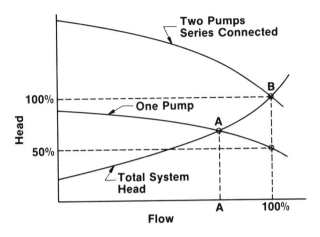

Figure 10-6. The effect of series pumping on the head curve.

system flow will occur at point B, which is the system design flow. In this operating mode both pumps are developing equal head at the full system flow.

In parallel pumping operations the pump component of total available head is identical for each pump, and the system flow is divided among the number of pumps operating in the system. The flows produced by individual pumps can represent any percentage of the total system flow. Where series pumping is described as "pressure additive," parallel pumping is described as "flow additive." When operating in parallel, pumps always develop an identical head value at whatever their equivalent flowrate is for that developed head, and the sum of their capacities will equal the system flow. Parallel pumping characteristics are illustrated in Figure 10-7.

In this example, each pump develops 50% of the total flow at 100% head. With one pump operating, the system flow will occur at point A; with both pumps in operation, flow will occur at point B.

High-Speed Coefficient

The high-speed coefficient (also referred to as the specific number of revolutions n_s) is the number of revolutions of geometrically similar models of an impeller, which at the same efficiency and capacity of 0.075 m³/sec has a head of 1 m. The high speed is a basic characteristic

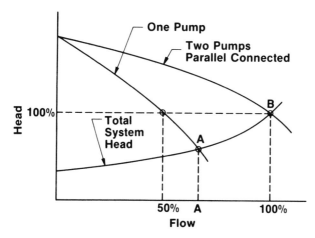

Figure 10-7. The effect of parallel pumping on the head curve.

of a series of similar pumps with equal angles α_2 and β_2, and coefficients ϵ and η_h. The high-speed coefficient is expressed by the following relationship:

$$n_s = \frac{3.65n\sqrt{Q}}{\sqrt[4]{H^3}} \tag{10-31}$$

where: n = rpm
        Q = maximum pump capacity, m³/sec
        H = total pump head, m.

The high-speed coefficient n_s increases with increasing capacity and decreasing head. Therefore, low-speed impellers are generally used for obtaining higher heads at low capacities, and high-speed impellers are used for creating high capacities at low heads. The impellers are divided into three groups depending on the value of the high-speed coefficient. Ranges of values are summarized as follows:

Type Speed	n_s
Low	40–80
Normal	80–150
High speed	150–300

Table 10-1
Major Pump Types and Construction Styles

Pump Type And Construction Style	Distinguishing Construction Characteristics	Usual Orientation	Usual No. Of Stages	Relative Maintenance Requirement	Comments
Dynamic					Capacity varies with head.
Centrifugal					Low to medium specific speed.
Horizontal					
Single stage overhung, process type	Impeller cantilevered beyond bearings.	Horizontal	1	Low	Most common style used in process services.
Two stage overhung, process type	2 impellers cantilevered beyond bearings.	"	2	Low	For heads above single stage capability.
Single stage impeller between bearings	Impeller between bearings; casing radially or axially split.	"	1	Low	For high flows to 330 m head
Chemical	Casting patterns designed with thin sections for high cost alloys; small sizes.	"	1	Medium	Low pressure and temperature ratings.
Slurry	Large flow passages, erosion control features.	"	1	High	Low speed; adjustable axial clearance.
Canned	Pump and motor enclosed in pressure shell; no stuffing box.	"	1	Low	Low head-capacity limits for models used in chemical services.
Multistaged, horizontally split casing	Nozzles usually in bottom half of casing.	"	Multi	Low	For moderate temperature-pressure ratings.
Multistage, barrel type	Outer casing confines inner stack of diaphragms.	"	Multi	Low	For high temperature-pressure ratings.
Vertical					
Single stage process type	Vertical orientation.	Vertical	1	Low	Style used primarily to exploit low NPSH requirement.
Multistage, process type	Many stages, low head/stage.	"	Multi	Medium	High head capability, low NPSH requirement.
In-line	Arranged for in-line installation, like a valve	"	1	Low	Allows low cost installation, simplified piping systems.
High Speed	Speeds to 380 m/s, head to 1770 m	"	1	Medium	Attractive cost for high head/low flow.
Sump	Casing immersed in sump for installation convenience and priming ease.	"	1	Low	Low cost installation.
Multistage deep well	Very long shafts	"	Multi	Medium	Water well service with driver at grade.
Axial (Propeller)	Propeller shaped impeller, usually large size.	Vertical	1	Low	A few applications in chemical plants and refineries.
Turbine (Regenerative)	Fluted impeller; flow path like screw around periphery.	Horizontal	1,2	Med. to High	Low flow-high head performance. Capacity virtually independent of head.
Positive Displacement					
Reciprocating					
Piston, plunger	Slow speeds; valves, cylinders, stuffing boxes subject to wear.	Horizontal	1	High	Driven by steam engine cylinders or motors through crankcases.
Metering	Small units with precision flow control system	"	1	Medium	Diaphragm and packed plunger types.
Diaphragm	No stuffing box; can be pneumatically or hydraulically actuated.	"	1	High	Used for chemical slurries; diaphragms prone to failure.
Rotary					
Screw	1, 2 or 3 screw rotors	"	1	Medium	For high viscosity, high flow high pressure.
Gear	Intermeshing gear wheels	"	1	Medium	For high viscosity, moderate pressure, moderate flow.

Miscellaneous Data

Tables 10-1 through 10-5 provide useful information and data on pumps. Table 10-1 provides a summary of major pump types and their construction characteristics. Table 10-2 provides a summary of operating limitations for different pump types. Table 10-3 provides convenient unit factors and formulas for centrifugal pumps. Table 10-4 provides typical suction head limits to avoid cavitation. Finally, Table 10-5 summarizes operating limits and applications of metering pumps.

Sample Calculation 10-1. A centrifugal pump transfers water from a storage tank to an elevated vessel at a rate of 12 m³/min. The pressure on the discharge side of the pump is 3.8 atm, and a vacuum gauge on the suction line reads 21 cm of Hg. The distance between the discharge side pressure gauge and vacuum gauge is 410 mm. The suction pipe diameter is 350 mm, and the diameter of the discharge line is 300 mm. Estimate the total head on the pump.

Solution. The water velocity in the suction line is

$$u_s = \frac{12}{60 \times 0.785 \times (0.35)^2} = 2.08 \text{ m/s}$$

The velocity in the discharge line is

$$u_d = \frac{12}{60 \times 0.785 \times (0.3)^2} = 2.83 \text{ m/s}$$

The pressure in the discharge line is

$$P_d = (3.8 + 1.03)10^4 = 48300 \text{ kg/m}^2$$

The pressure in the suction line is

$$P_s = (0.76 - 0.21)13{,}600 = 7500 \text{ kg/m}^2$$

The head on the pump is

(Text continued on page 176)

Table 10-2
Summary of Operating Performances of Pumps

Pump Type/Style	Solids Toler- ance	Capacity (dm^3/s)	Capacity (gph)	Max. Head (m)	Max. Head (ft)
Centrifugal					
Horizontal					
Single-stage overhung	MH	$1 \sim 320$	$950 \sim 3 \times 10^5$	150	492
2-stage overhung	MH	$1 \sim 75$	$950 \sim 7.1 \times 10^4$	425	1394
Single-stage impeller between bearings	MH	$1 \sim 2500$	$950 \sim 2.4 \times 10^6$	335	1099
Chemical	MH	65	6.2×10^4	73	239
Slurry	H	65	6.2×10^4	120	394
Canned	L	$0.1 \sim 1250$	$95 \sim 1.2 \times 10^6$	1500	4922
Multi. horiz. split	M	$1 \sim 700$	$950 \sim 6.7 \times 10^5$	1675	5495
Multi., barrel type	M	$1 \sim 550$	$950 \sim 5.2 \times 10^5$	1675	5495
Vertical					
Single-stage process	M	$1 \sim 650$	$950 \sim 6.2 \times 10^5$	245	804
Multistage	M	$1 \sim 5000$	$950 \sim 4.8 \times 10^6$	1830	6004
In-line	M	1-750	$950 \sim 7.1 \times 10^5$	215	705
High-speed	L	$0.3 \sim 25$	$285 \sim 2.4 \times 10^4$	1770	5807
Sump	MH	$1 \sim 45$	$950 \sim 4.3 \times 10^4$	60	197
Multi. deep well	M	$0.3 \sim 25$	$285 \sim 2.4 \times 10^4$	1830	6004
Axial (propeller)	H	$1 \sim 6500$	$950 \sim 6.2 \times 10^6$	12	39.4
Turbine (regenerative)	M	$0.1 \sim 125$	$95 \sim 1.2 \times 10^5$	760	2493
Positive Displacement				(kPa)	(psi)
Reciprocating					
Piston, plunger	M	$1 \sim 650$	$950 \sim 6.2 \times 10^5$	345000	50038
Metering	L	$0 \sim 1$	$0 \sim 950$	517000	74985
Diaphragm	L	$0.1 \sim 6$	$95 \sim 5.7 \times 10^3$	34500	5004
Rotary					
Screw	M	$0.1 \sim 125$	$95 \sim 1.2 \times 10^5$	20700	3002
Gear	M	$0.1 \sim 320$	$95 \sim 3.0 \times 10^5$	3400	493

Table 10-2 continued

Typical NPSH/Req.		Max. Kinematic Viscosity		Efficiency	Max. Pumping Temperature	
(m)	(ft)	(mm²/s)	(in.²/s)	(%)	(°C)	(°F)
2 ~ 6	6.6 ~ 20	650	1.01	20 ~ 80	455	851
2 ~ 6.7	6.6 ~ 22	430	0.67	20 ~ 75	455	851
2 ~ 7.6	6.6 ~ 25	650	1.01	30 ~ 90	205 ~ 455	401 ~ 851
1.2 ~ 6	3.9 ~ 20	650	1.01	20 ~ 75	205	401
1.5 ~ 7.6	4.9 ~ 25	650	1.01	20 ~ 80	455	851
2 ~ 6	6.6 ~ 20	430	0.67	20 ~ 70	540	1004
2 ~ 6	6.6 ~ 20	430	0.67	65 ~ 90	205 ~ 260	401 ~ 500
2 ~ 6	6.6 ~ 20	430	0.67	40 ~ 75	455	851
0.3 ~ 6	1 ~ 20	650	1.01	20 ~ 85	345	653
0.3 ~ 6	1 ~ 20	430	0.67	25 ~ 90	260	500
2 ~ 6	1 ~ 20	430	0.67	20 ~ 80	260	500
2.4 ~ 12	7.9 ~ 39.8	109	0.17	10 ~ 50	260	500
0.3 ~ 6.7	1 ~ 22	430	0.67	40 ~ 75	—	—
0.3 ~ 6	1 ~ 20	430	0.67	30–75	205	401
~ 2	6.6	650	1.01	65 ~ 85	65	149
2 ~ 2.5	6.6 ~ 8.2	109	0.17	55 ~ 85	120	248
3.7	12	1100	1.71	55 ~ 85	290	554
4.6	15.1	1100	1.71	~ 20	290	554
3.7	12.1	750 (ssu)	1.16 (ssu)	~ 20	260	500
~ 3	~ 9.8	150×10^6	150×10^6	50 ~ 80	260	500
~ 3	~ 9.8	150×10^6	150×10^6	50 ~ 80	345	653

MH—moderately high
H—high
M—medium
L—low

Table 10-3
Centrifugal Pump
Unit Factors and Formulas

CAPACITY

1 Cubic foot per second	= 449.	gpm
1 Million gallons per day	= 695.	gpm
1 Acre foot per day	= 449.	gpm
1 Litre per second	= 15.85	gpm

VOLUME

1 U.S. gallon	= 231.	Cu. Inches
	= 0.1337	Cu. feet
	= 3.785	Liters
	= 0.833	Imperial gal.
1 Imperial gallon	= 1.2	U.S. gallons
1 Cubic foot	= 7.48	U.S. gallons
	= 0.0283	Cu. meter
1 Liter	= 0.2642	U.S. gallons
1 Cubic meter	= 35.314	Cu. feet
	= 264.2	U.S. gallons
1 Acre foot	= 43,560	Cu. feet
	= 325,829	U.S. gallons

HEAD

1 Lb. per sq. inch	= 2.31	Ft. head of water
	= 2.04	Inches of mercury
	= 0.0703	Kg per sq. inch
1 Ft. of water	= 0.433	Lb. per sq. inch
1 Inch of mercury (or vacuum)	= 1.132	Ft. of water
1 Kilogram per sq cm	= 14.22	Lb. per sq. inch
1 Atmosphere	= 14.7	Lb. per sq. inch
	= 34.0	Ft. of water
	= 10.35	Meters of water

WEIGHT

1 U.S. gal. of water	= 8.33	Pounds
1 Cu. ft. of water	= 62.35	Pounds
1 Kilogram	= 2.2	Pounds
1 Metric ton	= 2204.6	Pounds

LENGTH

1 Mile	= 5280.	Feet
	= 1.61	Kilometers
1 Inch	= 2.54	Centimeters
1 Meter	= 3.2808	Feet
	= 39.3696	Inches

TEMPERATURE

Degrees Fahrenheit = $\frac{9}{5}$ Degrees Centigrade + 32

Degrees Centigrade = $\frac{5}{9}$ (Degrees Fahrenheit − 32)

PUMP PERFORMANCE WITH IMPELLER DIAMETER AND/OR SPEED CHANGE

Q_1, H_1, bhp_1, D_1 and N_1 = Initial Capacity, Head, Brake Horsepower, Diameter and Speed.

Q_2, H_2, bhp_2, D_2 and N_2 = New Capacity, Head, Brake Horsepower, Diameter and Speed.

DIAMETER CHANGE ONLY	SPEED CHANGE ONLY	DIAMETER & SPEED CHANGE
$Q_2 = Q_1 \left(\frac{D_2}{D_1}\right)$	$Q_2 = Q_1 \left(\frac{N_2}{N_1}\right)$	$Q_2 = Q_1 \left(\frac{D_2}{D_1} \times \frac{N_2}{N_1}\right)$
$H_2 = H_1 \left(\frac{D_2}{D_1}\right)^2$	$H_2 = H_1 \left(\frac{N_2}{N_1}\right)^2$	$H_2 = H_1 \left(\frac{D_2}{D_1} \times \frac{N_2}{N_1}\right)^2$
$bhp_2 = bhp_1 \left(\frac{D_2}{D_1}\right)^3$	$bhp_2 = bhp_1 \left(\frac{N_2}{N_1}\right)^3$	$bhp_2 = bhp_1 \left(\frac{D_2}{D_1} \times \frac{N_2}{N_1}\right)^3$

Formulas

$$gpm = .07 \times \text{Boiler HP}$$
$$gpm = 449 \times \text{cfs}$$
$$gpm = 0.0292 \times \text{BBL/Day}$$
$$gpm = 0.7 \times \text{BBL/Hour}$$
$$gpm = 4.4 \times \text{Cu. Meters/Hour}$$

$$gpm = \frac{\text{Lbs. per hour}}{500 \times \text{Sp. Gr.}}$$

$$H = \frac{P \times 2.31}{\text{Sp. Gr.}}$$

$$V = \frac{Q \times .321}{A}$$

$$U = \frac{\text{Diameter (inches)} \times N}{229}$$

$$h_v = \frac{v^2}{2g}$$

$$whp = \frac{Q \times H \times \text{Sp. Gr.}}{3960}$$

$$bhp = \frac{Q \times H \times \text{Sp. Gr.}}{3960 \times e}$$

$$bhp = \frac{Q \times P}{1715 \times e}$$

$$T = \frac{bhp \times 5250}{N}$$

$$N_s = \frac{N\sqrt{Q}}{H^{3/4}} = \frac{N\sqrt{H} \times \sqrt{Q}}{H}$$

$$S = \frac{N\sqrt{Q}}{h_{sv}^{3/4}} = \frac{N\sqrt{h_{sv}} \times \sqrt{Q}}{h_{sv}}$$

$$t_r = \frac{H\left(\frac{1}{e} - 1\right)}{780 \times C}$$

cfs	= Cubic Feet per Second
BBL	= Barrel (42 Gallons)
C	= Specific Heat
Sp. Gr.	= Specific Gravity
psi	= Pounds per Square Inch
gpm	= Gallons per Minute
e	= Pump Efficiency in Decimal
V	= Velocity in Feet per Second
D	= Impeller Diameter in Inches
T	= Torque in Foot Pounds
t	= Temp. in Degrees Fahrenheit
t_r	= Temp. Rise in Degrees Fahrenheit
A	= Area in Square Inches

N	= Speed in rpm
N_s	= Specific Speed in rpm
S_s	= Suction Specific Speed in rpm
Q	= Capacity in gpm
P	= Pressure in psi
H	= Total Head in Feet
h_{sv}	= Net Positive Suction Head in Feet
h_v	= Velocity Head in Feet
whp	= Water Horsepower
bhp	= Brake Horsepower
U	= Peripheral Velocity in Feet per Sec.
g	= 32.16 Feet per Sec. per Sec. (Acceleration of Gravity)
mgd	= Million Gallons per Day

Table 10-4
Typical Suction Head Limits to Avoid Cavitation*

Temperature (°C)	10	20	30	40	50	60	65
Suction head, (m)	6	5	4	3	2	1	0

Based on water flow.

Table 10-5
Operating Limits and Applications of Metering Pumps

Pump Type	Limits	Application	Wetted Parts Made Of
Low-pressure diaphragm pump with direct mechanical drive	<60-10 bar (85-140 psi), <500/h	Metering, pumping	PVC or austenitic stainless steel, elastomer diaphragm
Low-pressure bellows pump with direct mechanical drive	<5 bar (70 psi)	Metering, pumping	Glass; PTFE bellows
High-pressure micro-diaphragm metering pump with hydraulic drive	<700 bar (9,940 psi), <10/h	Metering	Acid-resistant steel
Diaphragm pump with hydraulic compression of tubular member	<50 bar (710 psi)	Metering, pumping	Acid-resistant steel; elastomer diaphragm and tube
Diaphragm pump with hydraulic drive	<350 bar (5,000 psi) for PTFE diaphragm, <3,000 bar (42,600 psi) for metal diaphragm	Metering, pumping	PVC, PTFE, titanium, acid-resistant steel

$$H = \frac{P_d - P_s}{\gamma} + H_0 + \frac{u_d^2 - u_s^2}{2g}$$

$$= \frac{48300 - 7500}{1000} + 0.41 + \frac{(2.83)^2 - (2.08)^2}{2 \times 9.81}$$

$$= 40.8 + 0.41 + 0.19$$

$$H = 41.4m \ H_2O$$

References

1. Cheremisinoff, N. P., *Fluid Flow: Pumps, Pipes and Channels,* Ann Arbor Science Publishers, Ann Arbor, MI (1981).
2. Cheremisinoff, N. P. and Azbel, D., *Fluid Mechanics and Unit Operations,* Ann Arbor Science Publishers, Ann Arbor, MI (1983).
3. Cheremisinoff N. P. and Gupta R. (editors), *Handbook of Fluids In Motion,* Ann Arbor Science Publishers, Ann Arbor, MI (1983).
4. Cheremisinoff, N. P., *Applied Fluid Flow Measurement,* Marcel Dekker Inc., New York, (1979).

GLOSSARY

Absolute Temperature. The Kelvin scale (the size of the degree is equivalent to the Centigrade degree) has all temperatures 273.16 degrees higher. In the Rankine scale, the size of the degree is the same as that of the Fahrenheit degree, but all temperatures are 459.69 degrees higher.

Accumulator. A vessel used for the temporary storage of a gas or liquid. Usually used for collecting sufficient material for a continuous charge to a downstream process.

Adiabatic Process. Process where there is no transfer of heat between system and surroundings.

Anhydrous. Free of water, especially water of crystallization.

API Gravity. An arbitrary scale denoting the gravity or density of liquid petroleum products. The measuring scale is calibrated in terms of "API" degrees. It can be computed from the following formula:

$$\text{deg. API} = \frac{141.5}{\text{sp.gr. } 60/60°F} - 131.5$$

Apparent Viscosity. Viscosity of a non-Newtonian material, defined as $\eta_a = \tau/\dot{\gamma}$, where τ is the shear stress and $\dot{\gamma}$ the shear rate. It is analogous to the Newtonian viscosity, μ.

Aspirator. Device which serves to create a partial vacuum through pumping a jet of water, steam, or some other fluid or gas past an orifice opening out of the chamber in which the vacuum is to be produced.

Atmosphere. A common unit of pressure, corresponding to the force of gravity acting on air above the earth's surface. A standard atmosphere is equal to 14.696 $lb_f/in.^2$, or 29.921 in. Hg at 0°C in a standard gravitational field.

Audigage. Nondestructive test instrument used to determine metal pipe and vessel wall thicknesses. The instrument measures thickness from one side only by using ultrasonic resonance to determine the fundamental frequency of the material being measured.

Baffle. Partial restriction, generally in the form of a plate, positioned to change the direction, guide the flow, or promote mixing within the vessel in which it is installed.

Ball Float. Device used inside a liquid storage vessel to sense the liquid level.

Ball Valve. A valve with a ball-shaped disc with a hole through the center of the ball, providing straight-through flow. A quarter-turn of the handle fully opens or closes the valve for quick shut off.

Bar. A unit of pressure in the SI system; equivalent to 0.9869 atm. or 10^6 $dyne/cm^2$.

Baumé Gravity. Arbitrary scale for measuring the density of liquids; the unit used is the "Baumé" degree. The scale uses an inverse ratio of the specific gravity scale:

$$\text{sp.gr.} = \frac{140°}{130°} = \text{Be degree}$$

This permits the translation of Baumé gravity to specific gravity. When floated in pure water, the Baumé hydrometer indicates 10°Be, while the specific gravity scale reads 1.00. The modulus 140 serves for liquids lighter than water, while the modulus of 145 is employed for liquids heavier than water. The Baume scale is employed by the Bureau of Standards for the measurement of all liquids except oils.

Bingham Fluid. Idealized flow behavior of a plastic fluid:

$$\tau = \tau_y + \eta \dot{\gamma}, \tau > \tau_y$$

$$\dot{\gamma} = 0, \tau < \tau_y$$

where τ_y is the yield stress and η is the plastic viscosity. The apparent viscosity of a Bingham fluid is:

$$\eta_a = \tau/\dot{\gamma} = \eta + \tau_y \, \gamma^{-1}$$

Bleeder Valve. A small valve installed in conjunction with a block valve to determine if the block valve is holding tightly. Also used to release pressure on a closed liquid or gas stream.

Block Valve. Valve used for isolation of equipment.

Blowcase. Closed vessel with product inlet and discharge lines and an air connection. With the manually operated type, air is admitted through the hand regulation of the valve to force the liquid out through the discharge pipe. With the automatic type, the flow of air is controlled by a float valve inside the vessel.

Bob. A calibrated weight on the end of a measuring tape used for gauging vessels.

Bourdon Gauge. Pressure indicator driven by the deformation of a curved tube of elastic metal to the interior of which the pressure is applied.

By-Pass Valve. Valve by which the flow of liquid or gas in a system may be directed past some part of the system through which it normally flows (e.g., an oil filter in a lubrication system).

Capacity Factor. Measure of the volume efficiency of an operation; generally expressed as the ratio of the actual daily throughput over a period of continuous operation to the demonstrated stream day capacity of a unit.

Centipoise. Unit of absolute viscosity. Viscosity of a Newtonian fluid is equal to one one-hundredth of the force in dynes which will move one centimeter per second relative to another parallel plane surface from which it is separated by a layer of fluid one centimeter thick.

Centistoke. Unit of kinematic viscosity; equal to centipoises divided by specific gravity.

Check Valve. Device for permitting flow in only one direction in a pipeline.

Control Valve. Variable opening valve used with a control instrument to maintain a predetermined flow rate, pressure, temperature, or level. Valve can be electric, electrohydraulic, or air-operated.

Critical Pressure. The pressure under which a substance may exist as a gas in equilibrium with the liquid at the critical temperature.

Critical Temperature. The temperature above which a gas cannot be liquefied by pressure alone.

de Florez Hole. A $1/8$-in. or $5/32$-in. diameter hole drilled in piping which is in corrosive service and which operates at high temperatures or is inaccessible for normal inspection measurements. The hole is normally drilled to a depth of $1/32$ in. less than the retiring wall thickness limit of the pipe. Leakage through a de Florez hole indicates thinning of the pipe to a point below the retiring thickness, thus providing a warning that maintenance is required.

Density. Mass of a fluid per unit volume:

$$\rho = \frac{m}{V} \left(\frac{kg}{m^3}\right)$$

Dew Point. The temperature at which liquid first condenses when a vapor is cooled.

Differential Pressure Flow Meter. Meter which senses the pressure drop across an orifice restriction which in turn is calibrated into the flowrate of the fluid.

Differential Pressure Transmitter. Device which measures differential pressure up to about 400 in. of water and converts these signals to pneumatic analogs which can be transmitted to remote reading instruments.

Dilatant Fluids. Also called shear-thickening fluids; fluids whose apparent viscosity increases with shear rate.

Ejector. Device which exploits the momentum of one fluid to move another. Also called a siphon, exhauster, jet pump, and eductor.

Expansion Joint. Joint used in the connection of long lines of pipe, which contains a bellows or telescope-like section to absorb the thrust or stress resulting from linear expansion or contraction of the line due to changes of temperature or to accidental forces.

Expansion Valve. A valve through which a fluid may be expanded from one pressure to a lower pressure at a controlled rate.

Flow Behavior Index. Properties parameter of a non-Newtonian fluid that can be described by a power-law model:

$$\tau = K\dot{\gamma}^n$$

where K is the consistency index and n is the flow behavior index. Both K and n are sensitive to temperature and pressure. For n $= 1$, the material is Newtonian; at n < 1, it is pseudoplastic or shear-thinning; at n > 1 it is dilatant or shear-thickening; τ and $\dot{\gamma}$ are the shear stress and rate, respectively.

Foot Valve. Check valve located at the inlet end of the suction line at a pump, which allows the pump to remain full of liquid when not in service.

Gaging. Refers to the act of measuring the height of liquid in a tank for the purpose of calculating the fluid inventory.

Gauge Pressure. Gauge or line pressure directly measured. Can be converted to absolute pressures by adding the barometric pressure.

Gate Valve. Valve with a sliding blank which opens to the complete cross section of the line; employed for complete opening or complete shutoff of the flow in pipes. It is not used for throttling or control.

Globe Valve. Valve used for throttling which does not have a straight-through opening.

Go-Devil. Cleaning device inserted in a pipe and forced through by air pressure or the movement of material in the pipe.

Hammer Testing. Method used by equipment inspectors to determine the presence of thin sections of piping, heads, and shells of pressure vessels, tanks, etc.

Heat Capacity. The quantity of heat required to raise the temperature of a given mass by $1°$. This quantity is based on either 1 mole or a unit mass of material.

Ideal Fluid. A fluid whose viscosity is zero and can thus be sheared without application of a shear stress.

Ideal Gas Law. Equation of state for an idealized gas $PV = nRT$, where $V =$ volume of n moles of gas, $P =$ absolute pressure, $T =$ absolute temperature, $R =$ univeral gas law constant.

Isobaric Process. A constant-pressure process.

Isometric Process. A constant-volume process.

Isothermal Process. A constant temperature process.

Kinetic Theory of Gases. Idealized model in which gases are considered to consist of minute, perfectly elastic particles which are moving at random with high velocities, colliding with each other and with the walls of the containing vessel. The pressure exerted by a gas is due to the combined effect of the impacts of the moving molecules upon the walls of the containing vessel, the magnitude of the pressure being dependent upon the kinetic energy of the molecules and their number.

Latent Heat of Vaporization. The heat required to convert a unit mass of material from the liquid state to the gaseous state at a given pressure and temperature.

Liquefied Gases. Normally applied to the liquid form of substances which under normal conditions of temperature and pressure are gases.

Look Box. Box with glass windows installed in a pipeline to permit visual inspection of the flow.

Manometer. U-tube measuring device for determining relatively low differential pressures.

Mechanical Seal. Mechanical device for sealing the flow of liquid along a centrifugal pump shaft by using a fixed and rotating element of two different materials.

Needle Valve. Valve with a cone seat and needle-point plug to control small and accurate flows.

Newtonian Fluid. Time-independent fluid that is described by the following constitutive equation $\tau = \mu\dot{\gamma}$, where τ is the shear stress, $\dot{\gamma}$ is the shear rate and μ the coefficient of consistency (i.e., viscosity). Viscosity is a function of temperature and pressure only for a Newtonian fluid.

Normal Boiling Point. The temperature at which a liquid boils when under a total pressure of one atmosphere.

Okadee Valve. Quick-opening valve used in liquefied gas service consisting of a handle-operating sliding disc that is held closed against a seat by the liquid pressure.

Orifice Plate. Steel plate with a sharp-edged circular restriction positioned in a pipe to measure flow by the differential pressure created across the restriction.

Pig. Jointed device that is forced through pipelines by hydraulic pressure to scrape off rust and scale or to mark an interface between two different products.

Plastic Fluids. Non-Newtonian fluids with a yield stress.

Plug Valve. Valve mainly used in gas service; consists of a rotating cylindrical plug in a cylindrical housing with an opening running through the slug.

Polytropic Process. No conditions other than reversibility are imposed on the system.

Positive Displacement Meter. Flow metering device which measures the total volume of liquid passing through it on a volume displacement basis.

Pressure. Defined as normal force exerted by the fluid per unit area of surface. Measured most frequently in terms of the height of a column of fluid under the influence of gravity. Column height measurements are converted to force per unit area by multiplying by the fluid's density $F_o/F = P = h\rho \, g/g_c$, where F is force.

Pseudoplastic Fluid. Referred to as a shear-thinning fluid, since the material's apparent viscosity decreases with shear rate.

Pump. Mechanical device for transporting liquids in pipelines. Major types are centrifugal, reciprocating, turbine, rotary, and proportioning pumps.

Reciprocating Pump. Positive displacement type pump using either pistons or plungers in the liquid cylinder end. Major types are simplex (1-cylinder), duplex, and multiples. Simplex and duplex units are most often steam or motor driven.

Relief Valve. Safety device for automatic release of fluid at a predetermined pressure.

Rheopectic Fluid. Time-dependent fluid whose shear stress at a constant shear rate increases slowly with time until an equilibrium value is reached. Such fluids behave as time-dependent dilatant fluids.

Rotameter. Variable area flow meter.

Rupture Disc. Device designed to fail at a specific pressure. It is used in lieu of a safety valve in installations where corrosion of the valves would be a problem and reseating is not essential.

Screw Pump. A rotary positive displacement-type pump applied when handling viscous liquids at high pressures above the range of the ordinary gear pump. The pumping rotors are meshing screws running at close clearance which progress from suction to discharge with trapped liquid to effect positive displacement.

Slide Valve. Valve comprised of a body with a large sliding disc, usually in the horizontal plane, which is actuated by compressed air or a hydraulic cylinder.

Specific Gravity. Weight of a unit of fluid volume:

$$\gamma = \frac{G}{V} \left(\frac{N}{m^3}\right)$$

Note that $\gamma = \rho g$, where g is the gravitational acceleration.

Specific Heat. The ratio of the heat capacity of a material to the heat capacity of an equal quantity of water. The specific heat of water is approximately 1 cal/(g)(°C). The molal heat capacity of water is 18 cal/(g-mol) (°C).

Specific Heat Ratio. Ratio of specific heat at constant pressure to the specific heat at a constant volume at a particular temperature.

Specific Volume. Volume per unit mass of fluid:

$$\dot{\nu} = \frac{V}{m} \left(\frac{m^3}{kg}\right)$$

That is $\dot{\nu} = 1/\rho$, where ρ is density.

Stilling Well. Pipe that is vertically installed in a vessel for gauging to eliminate splashing on the gauge when pumping into the tank.

Stuffing Box. Stationary enclosure through which a rotating shaft passes, designed to prevent fluid from leaking through. It is employed on pumps, agitators, etc.

Surface Tension. The work required for the formation of a unit of new surface:

$$\sigma = \left(\frac{J}{m^2}\right) = \left(\frac{N-m}{m^2}\right) = \left(\frac{N}{m}\right)$$

Surface tension can also be considered as a force acting on a unit length of the interface.

Temperature. A measure of a degree of hotness of a material detected most commonly with a liquid-in-glass thermometer. In the Centigrade or Celcius scale, the freezing point of water saturated with air at standard atmospheric pressure is zero; the boiling point of pure water at standard atmospheric pressure is 100. The Fahrenheit scale is t °F = 1.8 (t °C) + 32. The freezing point of water is 32°F and the boiling point is 212°F.

Thixotropic Fluid. Time-dependent fluid whose apparent viscosity decreases with increasing shear.

Time-Dependent Fluids. With such fluids, the sudden application of a change in shear rate results in the shear stress changing slowly with time until a new equilibrium shear stress is established corresponding to the changed shear rate. The shear stress is a function of shear rate and time until steady conditions are obtained. Two primary classes of time-dependent fluids are thixotropic and rheopectic.

Triple Point. The particular condition under which a substance can be present in any or all phases (gaseous, liquid, or solid).

Universal Gas Law Constant. Constant derived from the kinetic theory of gases in the ideal gas law. Values of the gas law constant R in different units are:

1.987 Btu/(lb-mol) (°F) or cal/(g-mol) (°K);
82.054 ml-atm/(g-mol)(°K);
8.3144×10^7ergs/(g-mol)(°K);
0.082054 liter-atm/(g-mol)(°K)

Ullage. Distance from the surface of the liquid in a vessel to the standard gauging point at the top of the tank. Also referred to as the outage.

Valve Positioner. Auxiliary servo device which allows precision positioning of a control valve stem. It is employed in conjunction with a standard valve operator (e.g., a diaphragm motor). Its purpose is to overcome stuffing box friction and stem thrust caused by fluid pressure.

Vapor Pressure. The pressure exerted by a vapor in equilibrium with the liquid phase of the same material.

Viscoelastic Fluids. Non-Newtonian materials that have properties analogous to elastic solids. They partially recover their original state after the stress is removed.

Viscosity. A measure of the tendency of a fluid to resist shear. The unit of viscosity is the poise which is defined as the resistance (in dynes per square centimeter of its surface) of one layer of fluid to the motion of a parallel layer one centimeter away and with a relative velocity of one centimeter per second.

Weir. Wall or partition for maintaining a liquid level.

Work. Product of force and the distance over which it is applied.

Yield-Power Law Model. Rheological model for describing plastic fluids which combines the Bingham and power-law models:

$$\tau = \tau_y + K\dot{\gamma}^n, \quad \tau > \tau_y$$
$$\dot{\gamma} = 0, \quad \tau < \tau_y$$

UNIT CONVERSION FACTORS

Table 1
Types of Base Units in the SI System

Symbol	Name	Quantity	In Terms of Other Units	In Terms of Base Units	Type of Unit
A	Ampere	Electric current	—	—	Base
a	Annum (year)	Time	365 d	$3.153\,600 \times 10^7$ s	Allowable
bar	Bar	Pressure	10^5 Pa	10^5 kg/(m·s^2)	Allowable
Bq	Becquerel	Activity (of a radionuclide)	—	$1\ s^{-1}$	Derived
C	Coulomb	Quantity of electricity, electric charge	—	1 A·s	Derived
°C	Degree Celsius	Celsius temperature	—	—	Derived
cd	Candela	Luminous intensity	—	—	Base
cP	Centipoise	Dynamic viscosity	1 mPa·s	10^{-3} kg/(m·s)	Allowable
cSt	Centistokes	Kinematic viscosity	1 mm^2/s	10^{-6} m^2/s	Allowable
d	Day	Time	24 hr	8.640×10^4 s	Allowable
F	Farad	Capacitance	1 C/V	$1\ s^4 \cdot A^2/(m^2 \cdot kg)$	Derived
g	Gram	Mass	—	10^{-3} kg	Allowable (submultiple of base unit)
Gy	Gray	Absorbed dose, specific energy imparted, kerma, absorbed dose index	1 J/kg	1 m^2/s^2	Derived
h	Hour	Time	60 min	3.6×10^3 s	Allowable
H	Henry	Inductance	1 Wb/A	$1\ m^2 \cdot kg/(s^2 \cdot A^2)$	Derived
ha	Hectacre	Area	—	10^4 m^2	Allowable
Hz	Hertz	Frequency (of periodic phenomenon)	—	$1\ s^{-1}$	Derived
J	Joule	Work, energy, quantity of heat	1 N·m	1 m^2·kg/s^2	Derived
K	Kelvin	Thermodynamic temperature	—	—	Base
kg	Kilogram	Mass	—	—	Base
kn	Knot	Velocity	1.852 km/h	$5.144\,444$ m^2/s	Allowable
L	Liter	Volume	1 dm^3	10^{-3} m^3	Allowable
lm	Lumen	Luminous flux	—	1 cd·sr	Derived
lx	Lux	Illumination	1 lm/m^2	1 cd·sr/m^2	Derived

Table 1 continued.

Symbol	Name	Quantity	Definition		Type of Unit
			In Terms of Other Units	In Terms of Base Units	
m	Meter	Length	—	—	Base
min	Minute	Time	—	60 s	Allowable
mol	Mole	Amount of substance	—	—	Base
N	Newton	Force	—	$1 \, m \cdot kg/s^2$	Derived
naut. mi	Nautical mile	Distance	1.852 km	$1.852 \times 10^3 \, m$	Allowable
Pa	Pascal	Pressure	$1 \, N/m^2$	$1 \, kg/(m \cdot s^2)$	Derived
r	Revolution	Angular displacement	$360°$	$2\pi \, rad$	Allowable
rad	Radian	Plane angle	—	—	Supplementary
S	Siemens	Conductance	$1 \, A/V$	$1 \, s^3 \cdot A^2/(m^2 \cdot kg)$	Derived
s	Second	Time	—	—	Base
sr	Steradian	Solid angle	—	—	Supplementary
Sv	Sievert	Dose equivalent	$1 \, J/kg$	$1 \, m^2/s^2$	Derived
t	Metric ton	Mass	1 Mg	$10^3 \, kg$	Allowable
T	Tesla	Magnetic flux density	$1 \, Wb/m^2$	$1 \, kg/(s^2 \cdot A)$	Derived
V	Volt	Electric potential, potential difference, electromotive force	$1 \, W/A$	$1 \, m^2 \cdot kg/(s^3 \cdot A)$	Derived
W	Watt	Power, radiant flux	$1 \, J/s$	$1 \, m^2 \cdot kg/s^3$	Derived
Wb	Weber	Magnetic flux	$1 \, V/s$	$1 \, m^2 \cdot kg/(s^2 \cdot A)$	Derived
Ω	Ohm	Electric resistance	$1 \, V/A$	$1 \, m^2 \cdot kg/(s^3 \cdot A^2)$	Derived
°	Degree	Plane angle	$(1/60)°$	$\pi/180 \, rad$	Allowable
'	Minute	Plane angle	$(1/60)°$	$2.908 \, 882 \times 10^{-4} \, rad$	Allowable
"	Second	Plane angle	$(1/60)'$	$4.848 \, 137 \times 10^{-6} \, rad$	Allowable

Table 2
Conversion Units Chart for Volumetric Flowrate

Given Units of:	Multiply by Table Values to Convert to These Units											
	m^3/s	dm^3/s	ft^3/d	ft^3/hr	ft^3/min	ft^3/s	U.K. gal/hr	U.S. gal/hr	U.K. gal/min	U.S. gal/min	bbl/d	bbl/hr
m^3/s	1	10^3	3.05119×10^6	1.2713×10^5	2.1189×10^3	3.5315×10^1	7.9189×10^5	9.5102×10^5	1.3198×10^4	1.5850×10^4	5.4344×10^5	2.2643×10^4
dm^3/s	10^{-3}	1	3.05119×10^3	1.2713×10^2	2.1189	3.5315×10^{-2}	7.9189×10^2	9.5102×10^2	1.3198×10^1	1.5850×10^1	5.4344×10^2	2.2643×10^1
ft^3/d	3.277×10^{-7}	3.277413×10^{-4}	1	4.1667×10^{-2}	6.9444×10^{-4}	1.15741×10^{-5}	3.7429×10^{-1}	3.1167×10^{-1}	6.2383×10^{-3}	5.1940×10^{-3}	1.781×10^{-1}	7.421×10^{-3}
ft^3/hr	7.866×10^{-6}	7.865791×10^{-3}	24	1	1.6667×10^{-2}	2.7778×10^{-4}	8.9831	7.48	1.4972×10^{-1}	1.2466×10^{-1}	4.274	1.781×10^{-1}
ft^3/min	4.719×10^{-4}	4.719474×10^{-1}	1.4400×10^3	60	1	1.6667×10^{-2}	5.3897×10^2	4.488×10^2	8.983	7.48	2.565×10^2	1.069×10^1
ft^3/s	2.832×10^{-2}	2.831685×10^1	8.6400×10^4	3600	60	1	3.234×10^4	2.693×10^4	5.3897×10^2	4.488×10^2	1.539×10^4	6.411×10^2
U.K. gal/hr	1.263×10^{-6}	1.262803×10^{-3}	2.6717	1.1132×10^{-1}	1.8554×10^{-3}	3.0923×10^{-5}	1	8.327×10^{-1}	1.667×10^{-2}	1.3878×10^{-2}	4.758×10^{-1}	1.983×10^{-2}
U.S. gal/hr	1.052×10^{-6}	1.051503×10^{-3}	3.20856	1.3369×10^{-1}	2.2282×10^{-3}	3.7136×10^{-5}	1.20094	1	2.00157×10^{-2}	1.667×10^{-2}	5.714×10^{-1}	2.381×10^{-2}
U.K. gal/min	7.577×10^{-5}	7.576820×10^{-2}	1.6030×10^2	6.6793	1.1132×10^{-1}	1.8554×10^{-3}	60	4.9961×10^1	1	8.3268×10^{-1}	2.855×10^1	1.189
U.S. gal/min	6.309×10^{-5}	6.309020×10^{-2}	1.9253×10^2	8.0220	1.337×10^{-1}	2.228×10^{-3}	7.2056×10^1	60	1.20094	1	3.428×10^1	1.429
bbl/d	1.840×10^{-6}	1.840131×10^{-3}	5.615	2.3396×10^{-1}	3.899×10^{-3}	6.499×10^{-5}	2.1017×10^1	1.750	3.503×10^{-2}	2.917×10^{-2}	1	4.1667×10^{-2}
bbl/hr	4.416×10^{-5}	4.416314×10^{-2}	1.3476×10^2	5.615	9.358×10^{-2}	1.5597×10^{-3}	5.044×10^1	42	8.407×10^{-1}	7.000×10^{-1}	24	1

Table 3
Conversion Units Chart for Force

Given Units Of:	Multiply by Table Values to Convert to These Units					
	g-cm-s^{-2} (dyne)	kg-m-s^{-2} (N)	lb$_m$-ft-s^{-2} (poundal)	lb$_f$	U.K. ton f	U.S. ton f
g-cm-s^{-2} (dyne)	1	10^{-5}	7.2330×10^{-5}	2.2481×10^{-6}	1.004×10^{-3}	1.124×10^{-3}
kg-m-s^{-2} (N)	10^5	1	7.2330	2.2481×10^{-1}	100.4	112.4
lb$_m$-ft-s^{-2} (poundal)	1.3826×10^4	1.3826×10^{-1}	1	3.1081×10^{-2}	1.388×10^1	1.554×10^1
lb$_f$	4.4482×10^5	4.4482	32.1740	1	4.464×10^2	5.00×10^2
U.K. ton f	9.964×10^2	9.964×10^3	7.207×10^{-2}	2.240×10^{-3}	1	1.120
U.S. ton f	8.896×10^2	8.896×10^3	6.435×10^{-2}	2.000×10^{-3}	0.8929	1

Table 4
Conversion Units Chart for Pressure

Multiply by Table Values to Convert to These Units

Given Units Of:	$g/cm\text{-}s^2$ $(dyne/cm^2)$	$kg/m\text{-}s^2$ (N/m^2)	$lb_m/ft\text{-}s^2$ $(poundal/ft^2)$	lb_f/ft^2	$lb_f/in.^2$ (psi)	Atmospheres (Atm)	mm Hg	in. Hg	bar	Pa	kPa
$g/cm\text{-}s^2$ $(dyne/cm^2)$	1	10^{-1}	6.7197×10^{-2}	2.0886×10^{-3}	1.4504×10^{-5}	9.8692×10^{-7}	7.5006×10^{-4}	2.9530×10^{-5}	10^{-6}	10^{-1}	10^{-4}
$kg/m\text{-}s^2$ (N/m^2)	10	1	6.7197×10^{-1}	2.0886×10^{-2}	1.4504×10^{-4}	9.8692×10^{-6}	7.5006×10^{-3}	2.9530×10^{-4}	10^{-5}	1	10^{-3}
$lb_m/ft\text{-}s^2$ $(poundal/ft^2)$	1.4882×10^{1}	1.4882	1	3.1081×10^{-2}	2.1584×10^{-4}	1.4687×10^{-5}	1.1162×10^{-2}	4.3945×10^{-4}	1.488×10^{-5}	1.488	1.488×10^{-3}
lb_f/ft^2	4.7880×10^{2}	4.7880×10^{1}	32.1740	1	6.9444×10^{-3}	4.7254×10^{-4}	3.5913×10^{-1}	1.4139×10^{-2}	4.78803×10^{-4}	4.78803×10^{1}	4.78803×10^{-2}
$lb_f/in.^2$	6.8947×10^{4}	6.8947×10^{3}	4.6330×10^{3}	144	1	6.8046×10^{-2}	5.1715×10^{1}	2.0360	6.89476×10^{-2}	6.89476×10^{3}	6.89476
Atmospheres (atm)	1.0133×10^{6}	1.0133×10^{5}	6.8087×10^{4}	2.1162×10^{3}	14.696	1	760	29.921	1.01325	1.01325×10^{5}	1.01325×10^{2}
mm Hg	1.3332×10^{3}	1.3332×10^{2}	8.9588×10^{1}	2.7845	1.9337×10^{-2}	1.3158×10^{-3}	1	3.9370×10^{-2}	1.33322×10^{-3}	1.33322×10^{2}	1.33322×10^{-1}
in. Hg	3.3864×10^{4}	3.3864×10^{3}	2.2756×10^{3}	7.0727×10^{1}	4.9116×10^{-1}	3.3421×10^{-2}	25.400	1	3.38638×10^{-2}	3.38638×10^{3}	3.38638
bar	10^{6}	10^{5}	6.720×10^{4}	2.088×10^{3}	1.450×10^{1}	9.869×10^{-1}	7.5006×10^{2}	2.953×10^{1}	1	10^{5}	100
Pa	10	1	6.720×10^{-1}	2.089×10^{-2}	1.450×10^{-4}	9.869×10^{-6}	7.5006×10^{-3}	2.953×10^{-4}	10^{-5}	1	10^{-3}
kPa	10^{4}	10^{3}	6.720×10^{2}	2.089×10^{1}	1.450×10^{-1}	9.869×10^{-3}	7.5006	2.953×10^{-1}	10^{-2}	10^{3}	1

Table 5
Conversion Units Chart for Viscosity

Multiply by Table Values to Convert to These Units

Given Units Of:	$g\text{-}cm^{-1}\text{-}s^{-1}$ (poise)	$kg\text{-}m^{-1}\text{-}s^{-1}$	$lb_m\text{-}ft^{-1}\text{-}s^{-1}$	$lb_f\text{-}s\text{-}ft^{-2}$	$lb_f\text{-}s\text{-}in.^{-2}$	centipoise	$lb_m\text{-}ft^{-1}\text{-}hr^{-1}$	$kg_f\text{-}s\text{-}m^{-2}$	mPa-s
$g\text{-}cm^{-1}\text{-}s^{-1}$ (poise)	1	10^{-1}	6.7197×10^{-2}	2.0886×10^{-3}	1.4504×10^{-5}	10^2	2.4191×10^2	1.0198×10^{-2}	10^2
$kg\text{-}m^{-1}\text{-}s^{-1}$	10	1	6.7197×10^{-1}	2.0886×10^{-2}	1.4504×10^{-4}	10^3	2.4191×10^3	1.020×10^{-1}	10^3
$lb_m\text{-}ft^{-1}\text{-}s^{-1}$	1.4882×10^1	1.4882	1	3.1081×10^{-2}	2.1584×10^{-4}	1.4882×10^3	3600	1.518×10^{-1}	1.4882×10^3
$lb_f\text{-}s\text{-}ft^{-2}$	4.7880×10^2	4.7880×10^1	32.1740	1	6.9444×10^{-3}	4.7880×10^4	1.1583×10^5	4.883	4.78803×10^4
$lb_f\text{-}s\text{-}in.^{-2}$	6.895×10^4	6.895×10^3	4.633×10^3	144	1	6.895×10^6	1.668×10^7	7.0309×10^2	6.89476×10^6
centipoise	10^{-2}	10^{-3}	6.7197×10^{-4}	2.0886×10^{-5}	1.4503×10^{-7}	1	2.4191	1.0198×10^{-4}	1
$lb_m\text{-}ft^{-1}\text{-}hr^{-1}$	4.1338×10^{-3}	4.1338×10^{-4}	2.7778×10^{-4}	8.6336×10^{-6}	5.995×10^{-8}	4.1338×10^{-1}	1	4.216×10^{-5}	4.1338×10^{-1}
$kg_f\text{-}s\text{-}m^{-2}$	9.806×10^1	9.806	6.589	2.048×10^{-1}	1.4223×10^{-3}	9.806×10^3	2.372×10^4	1	9.80665×10^3
mPa-s	10^{-2}	10^{-3}	6.719×10^{-4}	2.089×10^{-5}	1.4504×10^{-7}	1	2.419	1.0197×10^{-4}	1

Table 6
General Conversion Factors

To convert from ...	To ...	Multiply by ...
Atmospheres	Inches of mercury, at 0°C	29.92
Atmospheres	Pounds/square inch	14.70
Btu	Foot-pounds	778.3
Btu	Horsepower-hours	3.931×10^{-4}
Btu/hour	Foot-pounds/second	0.2162
Centimeters	Inches	0.3937
Cubic centimeters	Cubic feet	3.531×10^{-5}
Cubic centimeters	Cubic inches	0.06102
Cubic centimeters	Gallons (U.S. liquid)	2.642×10^{-4}
Cubic centimeters	Quarts (U.S. liquid)	1.057×10^{-3}
Cubic feet	Cubic centimeters	28320.0
Cubic feet	Cubic inches	1728.0
Cubic feet	Gallons (U.S. liquid)	7.48052
Cubic feet/minute	Cubic centimeters/second	472.0
Cubic feet/second	Gallons/minute	448.831
Cubic inches	Cubic centimeters	16.39
Degrees/second	Revolutions/minute	0.1667
Dynes	Kilograms	1.020×10^{-6}
Dynes	Pounds	2.248×10^{-6}
Feet of water	Inches of mercury	0.8826
Feet of water	Pounds/square inch	0.4335
Feet/minute	Centimeters/second	0.5080
Foot-pounds	Horsepower-hours	5.050×10^{-7}
Foot-pounds/minute	Foot-pounds/second	0.01667
Foot-pounds/second	Horsepower	1.818×10^{-3}
Gallons	Cubic centimeters	3785.0
Gallons	Cubic feet	0.1337
Gallons	Cubic inches	231.0
Gallons (U.S.)	Gallons (Imperial)	0.83267
Gallons/minute	Cubic feet/hour	8.0208
Horsepower	Foot-pounds/second	550.0
Inches	Centimeters	2.540
Inches of mercury	Kilograms/square centimeter	0.03453
Inches of mercury	Pounds/square inch	0.4912
Liters	Cubic inches	61.02
Liters	Gallons (U.S. liquid)	0.2642
Liters/minute	Gallons/second	4.403×10^{-3}
Millimeters	Inches	0.03937
Pounds/square inch	Atmospheres	0.06804
Pounds/square inch	Feet of water	2.307
Pounds/square inch	Inches of mercury	2.036
Pounds/square inch	Kilograms/square meter	703.1
Radians	Degrees	57.30
Radians	Minutes	3438.0
Radians	Seconds	2.063×10^{5}
Radians/second	Revolutions/minute	9.549
Radians/second	Revolutions/second	0.1592
Revolutions	Radians	6.283
Square centimeters	Square inches	0.1550
Square inches	Square centimeters	6.452
Square inches	Square millimeters	645.2
Square millimeters	Square inches*	1.550×10^{-3}
Temperature (°F) −32	Temperature, °C	5/9

INDEX